MONITORING and REMEDIATION WELLS

Problem Prevention, Maintenance, and Rehabilitation

Stuart A. Smith

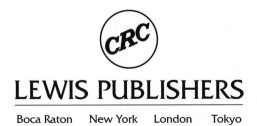

LEWIS PUBLISHERS

Boca Raton   New York   London   Tokyo

Library of Congress Cataloging-in-Publication Data

Smith, Stuart A.
  Monitoring and remediation wells : problem prevention,
  maintenance, and rehabilitation / Stuart A. Smith
      p.   cm.
  Includes bibliographical references and index.
  ISBN 0-87371-562-4 (alk. paper)
    1. Monitoring wells--Maintenance and repair.   2. Remediation
  wells--Maintenance and repair.   I. Title
  TD426.8.S65   1995
  628.1′61--dc20
                                                                                94-23958
                                                                                CIP

© 1995 by CRC Press, Inc.
Lewis Publishers is an imprint of CRC Press

No claim to original U.S. Government works
International Standard Book Number 0-87371-562-4
Library of Congress Card Number 94-23958
Printed in the United States of America 1 2 3 4 5 6 7 8 9 0
Printed on acid-free paper

Contents

Section II
Prevention and Maintenance

Section III
Rehabilitation and Reconstruction

List of Tables

List of Figures

The Author

Stuart Smith, CGWP, S.A. Smith Consulting Services in Ada, Ohio, U.S.A., conducts research and consults in the microbiology of groundwater and the troubleshooting and restoration of well problems (among other things). He otherwise leads the quiet life of a small-town Midwestern Lutheran family guy.

Stu holds B.A. and M.S. degrees from Wittenberg University (Ohio) and The Ohio State University, respectively. Prior to or along with consulting, he has worked for the National Water Well Association (long before it was the NGWA and there was an AGWSE) and taught water well technology at Wright State University in Ohio. He currently teaches biology at Ohio Northern University as well, and he advises student projects in hydrogeology for the ONU College of Engineering.

His interest in well iron biofouling began a while before he knew what it was — wondering about the slimy rusty stuff on pumps that he and his father pulled from wells in north-central Ohio. At NWWA in the early 1980s, Stu became the resident (and only functional) biologist, and he started working on "iron bacteria" in earnest. As a result, he became acquainted with the "motley crew" of microbiologists, hydrogeologists, drillers, and pump guys who are fascinated with this problem.

In the process of these various efforts, Stu has spoken to drillers, health officials, and environmental and agricultural audiences throughout North America and even in Australia and Argentina on well iron biofouling and well M&R. He also has managed to be the author or coauthor of over 100 publications (not all of them about iron biofouling), including most recently *Hydrofracturing for Well Stimulation and Geological Studies* (NGWA), the *Australian Drilling Manual* (ADITC), *Evaluation and Restoration of Water Supply Wells* and *Methods for Monitoring Iron and Manganese Biofouling in Water Supply Wells* (AWWARF), *Corrosión e incrustación microbiológica en sistemas de captación y conducción de agua* (Argentina Chamber of Commerce), and Well and Borehole Sealing (NGWA).

Stu is a member of NGWA (AGWSE), AWWA, the American Society for Microbiology, the Society for Technical Communication, ASTM, and the Ohio Academy of Sciences. He enjoys serving on the Ohio Section AWWA Small Systems Committee (you environmental people think you are the only ones who have OSHA and EPA problems...), the *Standard Methods* Section 9240 ("iron bacteria") task group, ASTM D-18 subcommittees relating to monitoring well development and restoration, and the U.K. CIRIA advisory committee for the development of a U.K.-specific well restoration guidance document.

In the past, he helped in the organization and operation of two symposia on aquifer and well biofouling and maintenance in Atlanta (1986) and at Cranfield Institute of Technology in the U.K. (1990). He also has an abiding interest in drilling training, but that is another story.

Preface
or
What Am I Supposed to Get from Reading This Book?

There is a growing problem of performance degradation of wells and associated systems on sites where groundwater quality is monitored or remediation is performed. It is getting to the point where people are beginning to doubt the validity of pump-and-treat as a method. In a lot of cases, the problem is not the concept, but the execution and the tough operating conditions these systems face.

This book is intended to be a guide in keeping monitoring and pumping well systems operating to their best capacity. It is written for those people who have to wrestle with such problems: site managers, their consultants and regulators, and contractors who may perform well and pump restoration services.

Problems experienced with monitoring and remediation wells are nothing new or unique. What site managers are experiencing has long before been a headache to a greater or lesser degree for operators of water supply, dewatering, recharge, and hydraulic-relief wells. Many of the same solutions apply in principle, but in a much more limited way in practice.

Monitoring and recovery wells are, after all, nothing more than specialized wells, often installed where no reasonable person would put a water supply well unless they were desperate. Recovery and treatment systems are nothing more than specialized groundwater-source water treatment systems. What does set them apart from a maintenance standpoint is that they are routinely exposed to environments and are operated in such a way that maximizes the potential for performance and water quality deterioration.

Even where groundwater is considered to be uncontaminated but is monitored due to potential hazards, monitoring wells are subject to greater deterioration effects than active pumped water supply wells, because they sit unused for long periods. Although such sentinel wells are monitored infrequently, their results must be reliable over long periods, perhaps ultimately over decades.

Evidence is mounting that fouling deposits filter out or absorb contaminants around monitoring and recovery wells. This filter effect potentially interferes with sampling and treatment. In-ground filtration may cause false negatives in samples, often resulting in erratic data over time. In some cases, fouling and bioaccumulation have caused the wells themselves to become point sources of contamination.

For these reasons alone (let alone performance), there is no room for hasty design, installation, development, or operation, no matter what the rationale might be for economizing or saving time on any of these aspects of project development and operation.

The process of operating any engineered system should include active maintenance. The alternative (in this case, the neglect of well and pump problems on environmental sites) leads to continued performance deficiencies, or even additional problems.

This book is intended to address the need for and methods of environmental well maintenance and restoration. It is a guidebook to the causes of well deterioration, methods of well maintenance, and well restoration and rehabilitation methods.

Like a useful travel guidebook, it is not a one-stop encyclopedia, but, where useful, points you to further sources of more information. In this case, the information for this work is built on the experience of the author and numerous other people, and a good chunk of that information is published elsewhere and should be on the bookshelf of—and read by—anyone responsible for environmental well systems. In particular, the American Water Works Association Research Foundation (AWWARF) has published two significant works, recommended to the reader, that pointed the way to the development of this book:

Evaluation and Restoration of Water Supply Wells. This is an encyclopedic review of the causes of well deterioration, methods for maintenance monitoring and treatment, methods for well rehabilitation and reconstruction, plus costs and related factors. Much of this also applies to environmental extraction and monitoring wells.

Methods for Monitoring Iron and Manganese Biofouling of Water Wells. Fe-related biofouling is the number one aquifer-related well deterioration problem. This report provides the conceptual basis for preventive monitoring of Fe and Mn biofouling, describes a field test program of such methods, and makes specific monitoring recommendations. Written for a municipal water supply system audience, the methods and decision-making are similar for and adaptable to environmental applications.

Another book complementary to this one is *Practical Groundwater Microbiology* (Lewis Publishers) by Roy Cullimore. This book describes in detail aquifer microbiological factors that cause well problems, and analytical and mitigation methods for microbial degradation of groundwater quality and well performance.

The work in front of you addresses the special needs of the environmental site: compliance, safety, and the peculiar technical needs of maintaining and rehabilitating these well systems.

The reader is recommended to these and other developing works coming from several sources. It was hard to stop and go to print this work, because the good stories keep rolling in; so keep up and keep informed, and by all means keep an open mind, seek all the good advice you can find, and respect it when you get it. In the words of actor Lavar Burton, host of PBS's *Reading Rainbow,* "You don't have to take my word for it."

Acknowledgments

The foresight and financial support of AWWARF, and the review and confidence of contributing AWWARF project advisory committee members, managers, reviewers, and my coauthors for the reports described in the Preface are gratefully acknowledged. Thanks to AWWA and AWWARF, modern literature finally exists that gives well operators and managers tools to plan and implement well maintenance and rehabilitation. I only wish we could have done more, because more needs to be done.

The input of a real innovator, practical guy, and survivor (not to mention a skilled storyteller) in well maintenance and restoration, George Alford, ARCC Inc., Daytona Beach, FL, is especially appreciated in this work. In addition, a variety of other consultants, contractors, and clients have provided illustrative lessons learned at a price, which are also gratefully acknowledged. The advice, information, and criticism provided by George, Olli Tuovinen, Roy Cullimore, Jay Lehr, and Peter Howsam over the years are particularly appreciated. Dr. Tuovinen's rigorous intellectual and writing standards have been a refining fire. Thanks also to the various reviewers of this manuscript, and Karen Ward, who helped with original art, for helping to tame this beast, which has been under development for more than 3 years.

Disclaimer

This work provides information on the problems of monitoring and remediation wells and their prevention and cures. It is not a substitute for experience. People reading this book should not consider themselves fully qualified to perform, specify, or supervise well maintenance and monitoring programs without guidance and experience with their specific situation. This should only be done by professionals experienced in well maintenance and rehabilitation, and who are qualified to work on hazardous and other environmental projects.

If you are a consumer of professional services in well rehabilitation, this book will help you to get the most from your professional help. If you are a provider, it is a source of information intended to help you do your job better and more safely. With that in mind, read on.

MONITORING and REMEDIATION WELLS

Problem Prevention, Maintenance, and Rehabilitation

I. Causes

Well deterioration is a serious concern in the operation of monitoring and plume control/remediation wells. Causes include formation, water quality, and biofouling, as well as operational factors.

1   A Brief Maintenance History of Environmental Wells

In order to understand and deal with well performance problems, it is necessary to understand some of the causes of well deterioration and how they affect the performance of the well. That's what the next two chapters do. Even if you now already have deteriorated wells and are looking for solutions, take some time to absorb this information.

THE HISTORY

Millions of wells have been constructed in the Industrialized World, mostly since the early 1980s, for a purpose other than the traditional ones: groundwater supply, recharge, or dewatering. Among these other purposes are monitoring groundwater quality and pumping to control or clean up contaminated groundwater.

At the same time that construction of such environmental wells has been accelerating, the "environmental industry" (consultants, government, drillers, and service users such as waste management firms) has been working to define and improve standards of performance in well construction, management, and use. The first order of business was to bring pump-and-treat, plume-control, and especially monitoring wells and systems, up to acceptable levels of performance.

The result has been the development of several important consensus standards, including ASTM standards for monitoring well construction and development. Some proposed standards, including one for maintenance and rehabilitation of monitoring wells, are still in draft form and are not yet adopted.

Improvements in execution have come from experience in the "field" and from training of drillers and supervisory people. An entire training and continuing education industry has sprung up to service the needs of professionals in the environmental industry so that monitoring and recovery systems could be competently designed and installed. Manuals on monitoring well construction and design (e.g., Aller et al., 1990 and Nielsen, 1991), not to mention improved methods, tools and equipment, and personnel skills, are part of the maturing of the industry. The results are not uniform—poorly designed systems are not disappearing—but we can hope that someday soon that day will come.

Maintenance and rehabilitation of these well systems, on the other hand, have not been adequately addressed (or even mentioned out loud much) in formal discourse until recently. In the literature, monitoring well maintenance and reha-bilitation are discussed in a preliminary way by Kraemer, Schultz, and Ashley (1991).

One reason for the lack of discourse may be that many existing wells were not worth reclaiming if their performance (pumping or water quality) declined. There were enough deficiencies in these existing monitoring and recovery wells that they were scheduled for replacement anyway. Maintaining them was not an issue.

As the industry gains experience with long-term monitoring and remediation systems, however, well maintenance does become more of an issue. Relatively well-designed and carefully constructed systems (judged by existing standards of performance) are still failing to perform up to expectations, and are experiencing a variety of symptoms. There is increasing direct and indirect evidence that well plugging and fouling of whatever type severely impact the performance of both monitoring and recovery wells. This calls into question assumptions about their performance.

In recovery and pump-and-treat systems, the chief problems are reduced flow and increased drawdown in the well systems and clogging of downstream piping and treatment apparatus. Pumps are a particularly hard-hit item (e.g., Hodder and Peck, 1992). Environmental well problems are fundamentally the same as those that cause water supply wells to provide poor performance. Poor design and poor construction and development can also contribute. However, inherent environ-mental causes of deterioration may occur even if design, installation, and devel-opment are adequate.

Monitoring wells may have less-obvious performance symptoms because they are not always stressed by pumping. Symptoms of well deterioration expe-rienced in monitoring wells are most likely to include changes in physicochemical water quality and increased turbidity. Such changes can interfere with the quality of samples from wells, as well as their performance. For example, Sevee and Maher (1990) describe a situation in which sediment in monitoring wells was interfering with the recovery of trichlorethylene, resulting in erratic TCE samples over time. Results became more consistent after wells were rehabilitated.

2   Causes and Effects of Performance Deterioration: The Lineup

This is the "rogues' gallery" of identified causes of poor well performance.

SUMMARY: CAUSES OF POOR WELL PERFORMANCE

There are numerous causes of poor and deteriorating well performance. Causes may include inherent characteristics of the formation (broadly speaking, an aquifer) that supplies water to the well, well design and construction, plus the groundwater quality. Operation of the well comes into play as well. Table 1 is a list of several categories of poor well performance or malfunction and likely causes.

SANDING AND SILTING

In a more ideal world, monitoring and recovery wells could be completed in favorable sand aquifer zones, and thus avoid many of the problems such wells have with clay, silt, and sand. The mission of such wells, however, is to be completed in discrete horizons to provide samples or contaminant recovery. For that reason, they are often completed in unfavorable zones, often at the top of the first "water bearing zone" or "aquifer" encountered.

Standard practice in monitoring wells is to complete screened wells with filter packs. These packs usually consist of uniform, rounded quartz sand of an average particle diameter suitable for holding out the more coarse material in a formation. However, selection of pack material and screen involves compromises, as has been pointed out repeatedly in numerous publications (e.g., Schalla and Walters, 1990; Rich and Beck, 1990; Nielsen and Schalla, 1991). The result is that the well screen and pack may not be suitable for retaining the finest material present in a screened interval. In rock wells, silting may result from an incomplete casing seat, which fails to seal off unconsolidated material.

It is also often difficult to properly place the filter pack material in the screen for optimal performance. Well annular spaces are usually small and not smooth. Bridging is highly likely. If the well is more than 40 ft or so, only the tremie

Table 1   Causes of Poor Well Performance

Sand/Silt Pumping

Inadequate screen and filter-pack selection or installation, incomplete development, screen corrosion, collapse of filter pack due to excessive vertical velocity and wash-out. Rock wells: presence of sand or silt in fractures intercepted by well completed open-hole, incomplete casing bottom seat. Causes pump and equipment wear and plugging.

Silt/Clay Infiltration

Generally inadequate seal around the well casing or casing bottom, infiltration through filter pack, or "mud seams" in rock, inadequate development, or overdevelopment in tills. Or material so fine that formation cannot be monitored without accepting some turbidity. Causes reduced performance, filter plugging, and interference with samples.

Pumping Water Level Decline

Outside influences such as area or regional water level declines or well interference, or plugging or incrustation of the borehole, screen, or gravel pack. Sometimes a regional decline will be exaggerated at a well due to plugging.

Lower (or Insufficient) Yield

Dewatering or caving in of a major fracture or other water-bearing zone, pump wear or malfunction, incrustation, plugging, or corrosion and perforation of column pipe, increased total dynamic head (TDH) in water delivery or treatment system.

Complete Loss of Production

Most typically pump failure (mechanical or electrical), but also possibly catastrophic loss of well production due to dewatering, plugging, or collapse.

Chemical Incrustation

Deposition of saturated dissolved solids, usually high Ca, Mg carbonate and sulfate salts or iron oxides. Causes reduced specific capacity and well efficiency, interference with sample analyses. Actually rare except for deep wells in highly mineralized groundwater, as in the U.S. West.

Biofouling Plugging

Microbial oxidation and precipitation of Fe, Mn, and S with associated growth and slime production. Often associated with simultaneous chemical incrustation and corrosion. Associated problem: well "filter effect": samples and pumped water are not necessarily representative of the aquifer. Usually includes "iron bacteria." Causes reduced specific capacity and efficiency, reduced yield, interference with sample quality, and even complete well production loss. Often works simultaneously with other problems such as silting.

Pump/Well Corrosion

Natural aggressive water quality, including H_2S, NaCl-type waters, biofouling and electrolysis due to stray currents. Aggravated by poor material selection in pump or column pipe, casing and screen. May result in secondary system symptoms.

Well Structural Failure

Tectonic ground shifting, ground subsidence, failure of unsupported casing in caves or due to poor grout support, casing or screen corrosion and collapse, casing insufficient for in-ground conditions, local site operations, collapse of unstable rock borehole.

Note: The Appendix includes a Prevention/Mitigation matrix of actions to prevent and limit deterioration (derived from Borch, Smith, and Noble, 1993) that visually simplifies this information.

methods can assure that pack material actually reaches the screen. Alternatively, a prepack method may be considered.

Beyond the inherent technical difficulties are the realities of field conditions. Numerous factors may interfere with exacting well installation. Drillers may be in a hurry because the consultants are pressing to meet a deadline or looking at a budget overrun. Supervisors of both drillers and field consultant personnel may want them to press the schedule because they have a backlog to work down. Site conditions may be poor, drilling equipment inadequate for the task (not uncommon), etc.

As is typically the case, drilling has commenced in the winter after the planning and approval cycle has been completed (having started after the beginning of the fiscal year). Everyone is cold, it is soggy, and work just does not go as well. The chances for a poor pack are enhanced. The fact that so much good work does get done is a credit to the real professionals of the industry.

The result, however, from a maintenance standpoint is that many wells have less-than-optimal screens, packs, and development. They are acceptable as monitoring points according to current criteria, but will pump some particulate matter. Under certain circumstances, especially in pumping wells, well clogging is likely and will have to be controlled so that the wells perform properly. Monitoring wells will produce turbid water that will have to be filtered for analysis, with unknown amounts of contaminants left adsorbed onto the suspended solids, thus affecting sample quality. Unless these are analyzed separately, sample fractions may be lost (e.g., Sevee and Maher, 1990).

YIELD AND DRAWDOWN PROBLEMS

In general, the same hydrologic considerations that are important to water supply wellfields are also important in monitoring and remediation well arrays. The hydrologic principles are, of course, the same.

Pumping water level decline or zone dewatering are typical symptoms in a pumping well. They may be symptoms of outside influences such as area or regional water level declines or well interference, or of reduced hydraulic efficiency in the well, resulting from plugging or encrustation of the borehole, screen, or filter pack. Typical causes of plugging are silt impacting on filter packs or biofouling (or usually both).

"Regional" declines in water tables or reduced storativity (due to compaction or biofouling precipitates filling pore spaces) contribute to pumping water level declines. They can also result in the dewatering of monitoring well intakes. When this occurs, of course, such wells become virtually useless as points for data gathering.

Dewatering of an aquifer zone can radically change the local biogeochemistry. Formerly reduced zones become oxidized, changing the nature of chemical constituents and the microbial ecology (Figure 1). This affects sample quality, because products being monitored may change. Contaminants may increasingly become attenuated.

Lower (or insufficient) yield may result either from hydrogeologic or mechanical causes. Dewatering or caving in of a major fracture or other water-bearing zone, or lack of connection to water-bearing zones, are primary geologic causes. Pump problems are much more common. They include pump wear or malfunction, encrustation, plugging, or corrosion and perforation of column pipe, and any increased total dynamic head (TDH) in the water delivery or treatment system.

Complete loss of production most typically results from pump mechanical or electrical failure. However, well mechanical causes may include catastrophic loss of well production due to dewatering, plugging, or collapse. Unless it is pump mechanical or power failure, usually there is some warning in the form of a noticeable well performance decline. Except for pump mechanical failure, complete well failure usually indicates some lapse or negligence in design, construction, or operation.

CHEMICAL ENCRUSTATION

Groundwaters typically maintain dissolved solids in solution, even when supersaturated. However, at the well, pressure and solubility changes may cause the deposition from saturated waters of Ca and Mg carbonate and sulfate salts. Redox and pH shifts may result in the deposition of iron oxides. Encrustation, as a problem separate from metal biofouling or other problems, is actually rare except for deep wells in highly mineralized groundwater, as in the U.S. West. Oxidation and changes in pH may be aggravated by high near-well turbulence and velocity, oxygen entrainment due to excessive drawdown, and microbial oxidation. Chemical encrustation may also be a secondary effect of biofouling oxidation or corrosion. Encrustation causes reduced specific capacity and efficiency, and interference with sample analyses.

CORROSION

Pump and well structural corrosion is a very complex phenomenon. Corrosion as a term is generally associated with metals and involves the removal of metal ions from metallic solids in contact with aqueous solutions. Causes of abiotic corrosion include naturally aggressive water quality, including those containing sulfides (H_2S or S^{2-}) and chlorides (Cl^-), and electrolysis due to stray electrical currents.

Corrosion occurs when the solution becomes ionized and water disassociated ($H_3O^+ + OH^-$) or there are other sources of excessive negative ions in the solution, such as chlorides (Cl^-). Positively charged metal ions become attracted to free hydroxyls (OH^-) or free anions such as Cl^-. Any corrosive situation is aggravated by poor material selection in the design of pump or column pipe, casing, and screen components. Corrosion has secondary effects such as sand pumping, alteration in water quality (especially elevated metals), secondary system clogging with corrosion products, and structural collapse.

PLASTIC DETERIORATION

Plastics are routinely used in wells and pumps used for sampling and pumping contaminated aquifers for very good reasons. Polyvinyl chloride (PVC), fluorocarbon (e.g., Teflon®), and polystyrene formulations used in casing and pump components are indeed recalcitrant (i.e., not susceptible to corrosion) by themselves.

PVC becomes subject to deterioration if it contains a significant percentage of plasticizers (Seal, 1990). PVC casing is normally rigid, with a low plasticizer content. Hydrocarbons are known to penetrate PVC pipe, however, and may serve as a means of softening PVC bonds and making its polymer components available for biodegradation in some circumstances. While solvent cements are not used in assembling monitoring well casing or plastic piping in these systems, it has been the practice in some places to solvent-weld casings for remediation pumping wells. These cement bonds are softened and broken in some organic-laden ground-water environments.

WELL STRUCTURAL FAILURE

Catastrophic structural failure is relatively rare in water supply wells, but apparently not so rare on remediation sites. Causes include regional forces such as tectonic ground shifting and ground subsidence (usually resulting from overpumping). Well construction failures may result from unsupported casing in caves or due to poor grout support, casing or screen corrosion and collapse, casing insufficiently strong for in-ground conditions, screen collapse due to prolonged sand-pumping, and the collapse of unstable rock boreholes. Vehicles are a major cause of monitoring well damage. Collisions cause casing breaks and dislocation.

Total structural failure results in many of the same symptoms as in pump and casing or screen corrosion (Figure 2). Turbidity or sand pumping rather suddenly increases, and yield may dramatically decline (if the pump remains functional). Related problems include pump and system wear and pump corrosion.

BIOFOULING: A BIG BUG PROBLEM

Environmental well systems are usually designed by geologists and engineers from a geological point of view. Naturally, plugging by sediment (sand, silt, clay) is the most-often recognized cause of well plugging and reduced performance (e.g., Kraemer, Schultz, and Ashley, 1991). Due to the nature of many monitored or pumped "aquifers," such mechanical plugging due to silt and clay is indeed a major factor.

However, experience with a wide range of well types (water supply, dewatering, recovery, and monitoring) suggests that the number one contributor to reduced well performance in most regions is biofouling. Where silting is indicated as the plugging cause, it is most likely working in tandem with biofouling plugging at or near the intake surface and the portion of the aquifer matrix subject to partial oxidation.

Biofouling involves the biological formation and deposition of fouling materials, which usually include mineral and metal precipitates (Fe, Mn, or S). Because of its importance and relative unfamiliarity as a problem to many, biofouling will be discussed here at length.

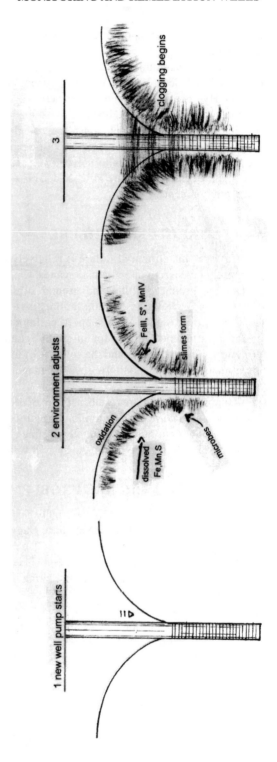

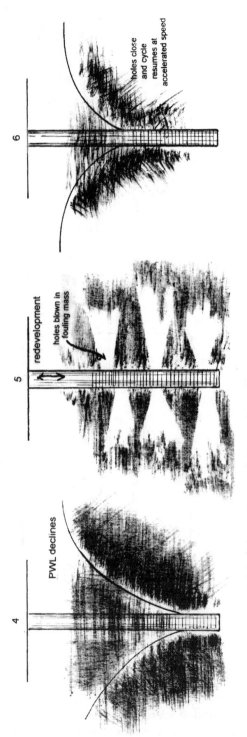

Figure 1.   Causes of accelerated pumping water level decline.

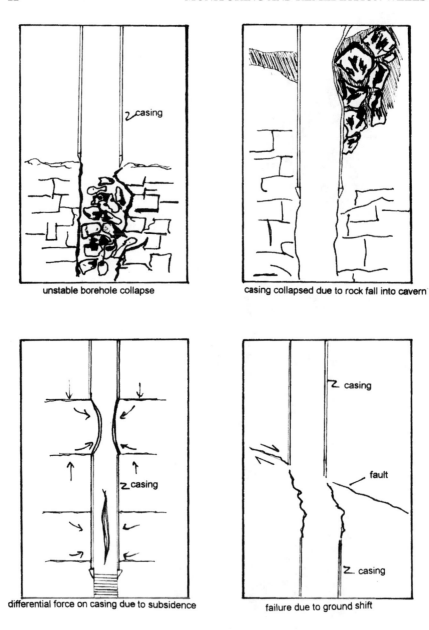

Figure 2.   Causes of well structural failure.

Biofilm and Biofouling Basics

Biofouling can take many forms in engineered systems, from zebra mussel clogging of surface water intakes to fairly benign coatings in all sorts of industrial systems, both on the surface and underground. Such complex biological coatings are known as biofilms. Zebra mussels are just really chunky biofilms. Figure 3 illustrates some manifestations of biofilm formation.

Biofilms and Microbial Survival

Biofilms have a survival function for microorganisms in aquatic environments. One is to provide, on a microscopic scale, multiple environments within the biofilm, allowing for the survival of a variety of microorganisms, transport of nutrients, and physicochemical gradients. Such microenvironments cannot be evaluated in tests of bulk groundwater quality.

Biofilms also protect cells within them from external stress such as disinfectants. Fe-, S-, and Mn-precipitating bacteria in these biofilms precipitate metallic oxides and extracellular polymers (ECP). Anaerobic bacteria associated with biofilms produce reducing agents (e.g., sulfides and methane). All of these can react with oxidants such as chlorine, oxygen, and hydrogen peroxide. Buffers generated by the bacteria can likewise dampen the effects of acids or caustics used in well treatments.

Biofilms and Biofouling in Groundwater

It should hardly be surprising to an industry accustomed to *in situ* bioremediation that aquifers often contain large, active microbial populations. In groundwater source systems, microbial biofilms are the predominant habitat for aquifer microflora. From a microbial standpoint, aquifers are ideal environments in many cases, especially if suitable organic substrates are present. There is tremendous surface area for colonization, moderate temperatures, nutrient flux, and overall very little disturbance. Inasmuch as the surface areas of the interstitial spaces in an aquifer are very large, the total mass of biofilms around a well can be likewise large given the right conditions. These biofilms are the basis for biologically mediated encrustation in wells and associated corrosion.

Where significant amounts of hydrocarbons and other organic compounds are mixed with water, large populations of the many microorganisms that can utilize these compounds also occur. This is particularly the case where *in situ* bioremediation is encouraged by the addition of microbial nutrients, cometabolites, and electron acceptors (PO_4, NO_3, O_2, etc.).

A widely observed phenomenon in ecology is that community diversity and total live biomass is greatest where there are gradients, such as forest edges. In the microbial realm, these gradients are primarily physicochemical in nature and can occur over microscopic distances.

A variety of environmental gradients can be expected to be formed by biogeochemical processes in aquifers. Such processes are enhanced in aquifers with organic concentrations. These may be sufficient to encourage microbial growth and sharp changes in redox potential. These gradients facilitate (or are the product of) a variety of microbial activities. This variety is reflected in a high overall microbial diversity in relatively "rich" groundwater, although single species may dominate locally. Fermentation, chemoheterotrophic oxidation of organics, and both oxidation and reduction of minerals and metals are practiced by microorganisms all living in close proximity.

More hydrophobic organics may adhere to particles where the biofilms form so that the bacteria present have an especially favorable situation for mass growth.

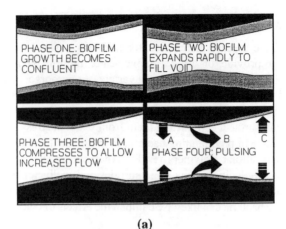

(a)

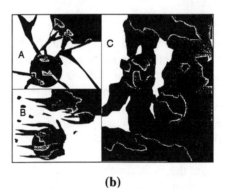

(b)

**Figure 3.   Biofilm schematics. (Cullimore, 1993.) (a) Four phases of biofilm
growth; (b) forms of interstitial biofilm growth; and (c, d) configuration of biofouling deposits around a pumping well and screen.**

Hydrocarbon-water interfaces provide a favorable growth situation as well (van
Loosdrecht et al., 1990). On a soil particle coated with dilute product, both
situations occur (solid and oil-water interfaces), providing more variety in growth
conditions (Figure 4).

The microbial ecology of a remediation system may be quite complex
(Figure 5), with molds occupying the vadose zone, anaerobic dissimilatory
metal reducers occupying the anaerobic groundwater, and metal oxidizers and
hydrocarbon utilizers found mostly at the interface between the aerobic and
anaerobic zones.

Water Quality Degradation: Monitoring and Remediation Problems

Typically, the earliest noticeable manifestation of biofouling in any well is
the water quality change that accompanies mass bacterial growth and associated
Fe, Mn, and S transformations (see the following). Turbidity increases and filters

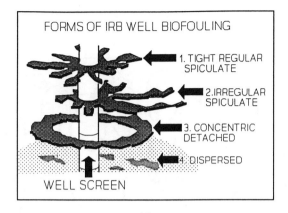

FORMS OF IRB WELL BIOFOULING

1. TIGHT REGULAR SPICULATE

2. IRREGULAR SPICULATE

3. CONCENTRIC DETACHED

4. DISPERSED

WELL SCREEN

(c)

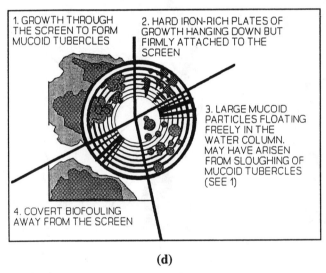

1. GROWTH THROUGH THE SCREEN TO FORM MUCOID TUBERCLES

2. HARD IRON-RICH PLATES OF GROWTH HANGING DOWN BUT FIRMLY ATTACHED TO THE SCREEN

3. LARGE MUCOID PARTICLES FLOATING FREELY IN THE WATER COLUMN. MAY HAVE ARISEN FROM SLOUGHING OF MUCOID TUBERCLES (SEE 1)

4. COVERT BIOFOULING AWAY FROM THE SCREEN

(d)

Figure 3.   (Continued.)

are clogged, or filtration of samples becomes necessary. Odors may change or form — which are further symptoms of microbial processes.

Discoloration, high bacterial counts, high turbidity (excluding sediment), and odor are symptomatic of active, established biofilms present in and around the pumping well. Portions of biofilms will intermittently slough off into the water being pumped through the collection-distribution system to the treatment plant.

Transient, elevated Fe, Mn, and H_2S concentrations in pumped groundwater, and increases in levels at the front of a plume, are typically the result of bacterial activities in the aquifer, including metabolic Fe(III), Mn(IV), and SO_4^{2-} reduction mobilizing soluble Fe, Mn, and S species. Subsequent sloughing of Fe(III) and Mn(IV) biofilms containing sulfide products add to total metal contents. Figure 6 is a representation of Fe, Mn, and S transformations in aquifers.

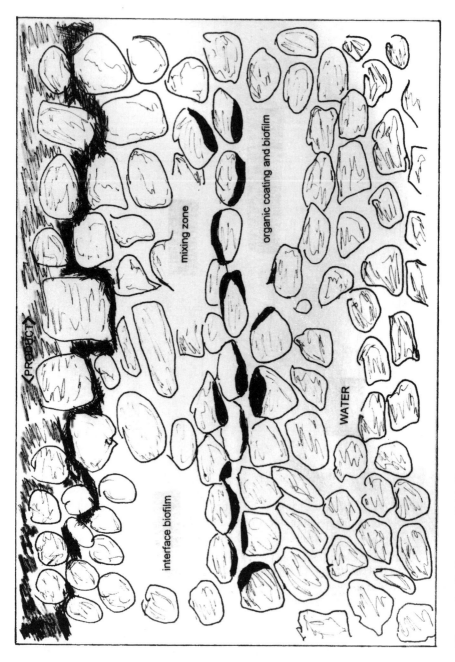

Figure 4.   Soil-water-oil-biofilm schematic.

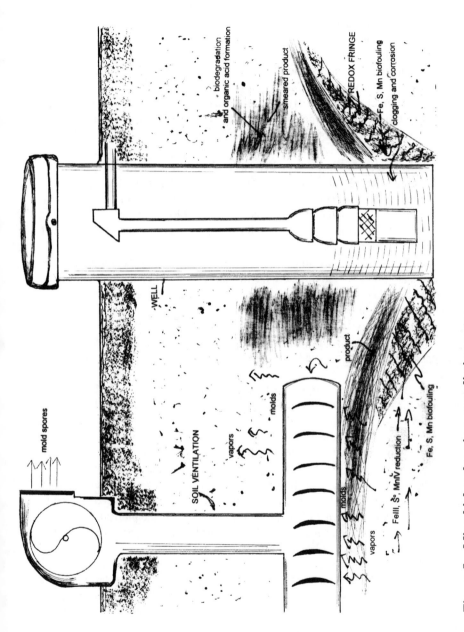

Figure 5. Microbial ecology in a remediation system.

Metallic oxides (predominantly Fe hydroxides) produced by corrosion and Fe(III) precipitation due to biofouling are important reactive surfaces. They interact with charged species such as H^+ (thus affecting pH), Cd^{2+} and other metallic cations, and anions such as SO_4^{2-}, as well as organic compounds (van Riemsdijk and Hiemstra, 1993; Davis et al., 1993). Mn(IV) and Fe(III) oxides can thus scavenge heavy metals such as Co, Ni, Cu, Zn, and Sn, for example, as described by Vuorinen and Carlson (1985). Fe hydroxides are also involved in the immobilization of soluble U(VI), e.g., as described by Waite and Payne (1993).

Many types of microorganisms can selectively precipitate minerals. Lovley (1991) describes the role of microbial reduction in forming insoluble U(IV) from soluble U(VI), potentially immobilizing uranium in the subsurface. In a similar case, *Pedomicrobium* species have been identified as facilitating the capture and reduction of gold chloride to elemental gold in soils (Alper, 1992).

Levels of organic constituents detected in monitoring well samples may become erratic over time (e.g., Sevee and Maher, 1990). This may be the result of partial attenuation on soil particles, followed by release in sloughing events, but biofilms may also cause partial attenuation and sloughing. Fe hydroxide minerals associated with biofilms are highly reactive with reduced organic compounds. The cycle of adsorption, partial utilization, and slug release results in sample interference and data that indicate irregular concentrations, or show breakdown products and "unknowns" in organic analyses.

Fluctuating raw water Fe and Mn levels are typical of biofouled wells, including monitoring wells. Mn and Fe complexed with ECP may occur in suspension and can be detected in high levels in analytical results from unfiltered samples.

Bacterial ECP that sequester metal ions are involved in Fe and Mn mobility in a process analogous to Fe sequestration treatments used to control Fe and Mn precipitation in water distribution systems. Lovley (1991) provides a detailed description of the sequences of Fe(III) and Mn(IV) reduction and Fe(II) and Mn(II) mobilization that result in dissolved Fe^{2+} and Mn^{2+} in groundwater.

Conversely, biofilms may also act as metal filters, removing Fe and possibly Mn from solution. Once the well biofilm has been removed or inactivated during rehabilitation treatment, Fe and Mn levels may increase in the produced water due to the lack of a sequestering effect previously provided by the biofilm. Cullimore (1993) has adapted the terms "causal" and "postdiluvial" water to describe the results of this commonly observed process (Figure 7).

The causal water is the representative groundwater of interest, relatively unaffected by effects associated with the well (biofouling and oxidation due to drawdown), although subject to alteration due to redox conditions in the "natural" aquifer. The postdiluvial water is that which is carefully collected and analyzed at the well. If collected properly, such samples are assumed to be representative, but may actually be products of microbial and other well-associated redox modification.

Intermittently high Fe and Mn levels can have a considerable effect on the operation of aquifer remediation treatment plants. These effects include in-

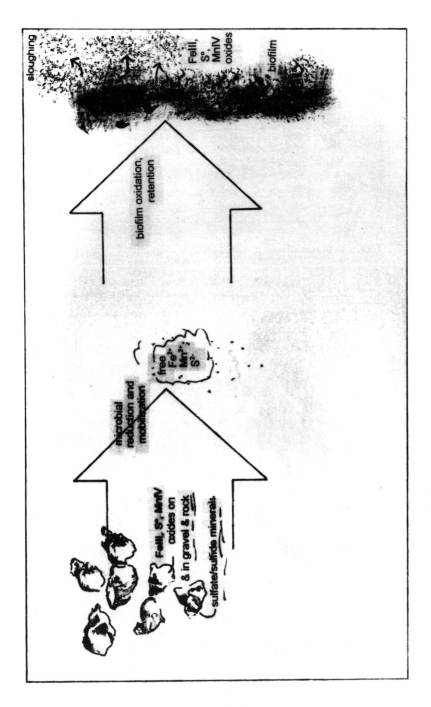

Figure 6. Fe, Mn, and S transformations and mobility in aquifers.

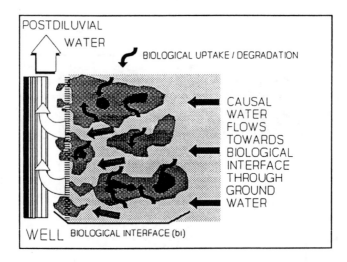

Figure 7.   Concept of causal and postdiluvial water quality. (Cullimore, 1993.)

creased chemical oxidant demand, bleed-through of filters, increased system hydraulic head, and more frequent filter backwash and aeration tower cleaning intervals.

These biologically mediated water quality changes can be particularly troubling when Fe(III)-oxide or S-slime deposits clog pipelines (typically not very large in diameter anyway) and coat carbon filters or aeration tower media. It is clear that microbial effects can be very important influences on monitoring and recovery wells, just as they are for water wells; and dealing with these effects needs to be a part of design and operation.

Microbially Mediated Metallic Corrosion

Microorganisms accelerate corrosion of well and water treatment system metallic components in a variety of ways. The microbial enhancement of corrosion is related to the production of corrosive metabolites such as organic acids (sources of metal scavenging R-COO$^-$ anions) and sulfides (S^{2-}), as well as the establishment of differential-oxidation cells in the form of colonies or biofilms. The formation of these cells creates conditions for anodic dissolution of metals, particularly iron.

Likewise, Fe(III)-reducing bacteria have been demonstrated to reduce Fe-oxide coatings on steel, stimulating corrosion (Lovley, 1991). Fe biofouling is typically not uniform, and areas of differing electron potentials result on biofouled surfaces, providing local environments conducive to corrosion.

When differential-oxidation cells form, electrochemical corrosion occurs at areas (anodes) of lowest oxygen concentration, for example, under a biofilm. In addition to providing a diffusion barrier to oxygen mass transfer, the microorganisms present in biofilms consume available oxygen by aerobic respiration. Figure 8 is a schematic of microbial corrosion processes.

When oxygen depletion occurs, anaerobes such as sulfate (SO_4)-reducing bacteria (SRB) can proliferate if suitable organic carbon sources are present. They seem to be virtually ubiquitous in aquifers, even those with overall high bulk oxidation-reduction potential (Eh). SRB can utilize molecular hydrogen and produce S^{2-}, both of which are important in electrochemical corrosion. In addition, biofilms shelter other heterotrophic bacteria that produce acidic metabolites which are corrosive or serve to promote or maintain reducing environments that benefit the SRB. One example is the production of short-chain organic acids during incomplete anaerobic oxidation of long-chain aliphatic hydrocarbons.

Such metal corrosion processes are accelerated in recovery and purge well systems controlling organic plumes, due to both the presence of organics that are degraded to organic acids and the overall intensity of microbial activity. This microflora may additionally include a wide range of fungal growth and functions where wells and sediments are frequently unsaturated. Thus, there is a complex overall biochemical situation that encourages multiple electron-potential cells and corrosion.

The most typical expression of metallic corrosion in environmental monitoring and pumping wells is intergranular corrosion cracking of stainless steel. Corrosion initiates in heat-affected zones such as weld areas of wire-wound screens at features, called nodes, and spreads. Nodes may have small external openings but large subsurface expression. Heated austenitic steel is sensitized at 950 to 1450°C, with chromium depletion at intergranular boundaries.

Biofilms interfere with the oxidation at the metallic surface necessary for repassivation of the stainless steel by forming differential oxidation cells. Biofouling proceeds as in mild steel (Pope et al., 1989).

Iron, Manganese, and Sulfur Biofouling

Fe, Mn, and S biofouling can be considered as a particular case in the biofouling topic, primarily because of its prevalence and impact on the performance of well systems and downstream piping and treatment. Such biofouling is complex and is a factor in the water quality and corrosion effects just discussed.

Fe, Mn, and S Biofouling: What's Happening

Microbiological activities are now regarded as the most important factor in the oxidation-reduction reactions that take place in groundwater, both for inorganics and many organics, even under anaerobic conditions (Chapelle, 1992). Bacteria are able to utilize substrates such as hydrocarbons in the absence of oxygen by using other electron acceptors such as Fe^{3+}, SO_4, and NO_3 (e.g., Lovley and Lonergan, 1990; Riss and Schweisfurth, 1985; Beller, Grbic-Galic, and Reinhard, 1992). Microbially mediated redox reactions can be complex and add greatly to the problem of fully understanding the geochemical environment in an aquifer setting (Figure 9).

Shallow aquifers rarely have such low redox potentials that organic oxidation reactions are not favorable, unless the organic content and microbial activity are

Figure 8.   Microbial corrosion aspects of biofouling.

quite high and oxygen is entirely depleted. This, of course, is likely to be the case when hydrocarbon contamination occurs.

In groundwater source systems, including monitoring and recovery wells, biofouling usually involves the oxidation of Fe, Mn, and S compounds by bacteria. These compounds become part of biofilm complexes including the Fe, Mn, and S compounds, ECP, and the bacterial cells themselves.

Fe and Mn biofouling can vary from being a minor nuisance to a cause of major maintenance problems, even resulting in complete abandonment of wells and wellfields. Fe and Mn biofouling problems are well documented with numerous reports published in a variety of water supply and groundwater industry literature based on experience from North America and around the world.

S-slime biofouling is much less well documented, but is typical of wells operating in sulfide-containing groundwaters (including those recovering hydrocarbons) where oxidizing conditions exist in pumping wells. Problems include clogging of pumps, screens, and filters by slimes and subsequent precipitates such as Ca sulfate, and associated corrosion of metal pump components.

Microbially mediated Fe, Mn, and S transformations also play an integral role in fouling attached water collection and treatment systems. As in wells, corrosion, the formation of encrustation and slimes, and changes in Fe, Mn, and S content and form in pumped water, occur downstream.

How Fe, Mn, and S Biofouling Occurs

Fe and Mn biofouling may take many forms and may be caused by both direct and indirect or passive microbial processes. Fe(II) compounds (ferrous Fe) or ions (Fe^{2+}) can also be oxidized to the Fe(III) (ferric) state by nonbiological (abiotic) oxidants such as chlorine or oxygen. The precipitation of Fe and Mn due to bacterial action is often hard to distinguish from abiotic processes. However, in aquifers with typically high (>0.1 mg/l) total organic carbon and bulk Eh-pH conditions in the stability range for dissolved species of Fe and Mn, abiotic processes are unlikely to be the immediate cause of Fe(III) and Mn(IV) oxide precipitation. The more likely cause is microbial.

Microbial oxidation of Fe and Mn: Some bacterial strains associated with iron biofouling have been shown to actively (enzymatically) oxidize Fe and Mn for various purposes (Emerson and Ghiorse, 1992; Corstjens et al., 1992; Hallbeck, 1993), and others are suspected, especially the common "iron bacterium," *Gallionella ferruginea.* Besides direct enzymatic oxidation, Fe(III) and sometimes Mn(IV) oxidation is also favorably mediated by microbial structures and reaction with ECP under common groundwater conditions (e.g., Gounot and di Ruggiero, 1990; Mouchet, 1992).

Microbial S oxidation: Microbial oxidation of sulfides results in the familiar white slime phenomenon of sulfur springs. Sulfur-oxidizing bacteria (SOB) have been found to be relatively common in aquifer sediments (Fredrickson et al., 1989) and also in wells developed in S^{2-}-containing groundwaters (Smith, 1991).

S^0 has a narrow stability range (Eh approx. 200 to 400 mV vs. pH 4 to 2) under the very specific environmental conditions of Hem's (1985) stability diagrams.

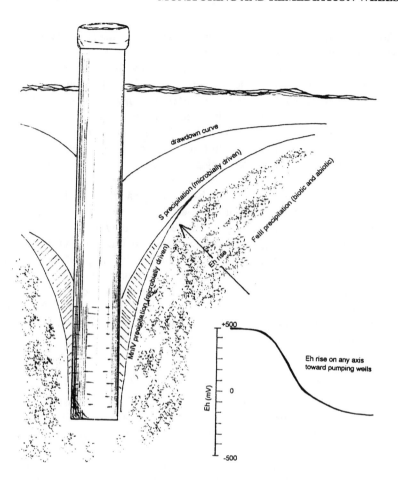

Figure 9. Redox fringe schematic for Fe, Mn, and S.

Bacteria precipitating S^0 apparently provide this environmental condition in their biofilms for as yet unknown reasons.

The expression of S biofouling resembles Fe and Mn biofouling in that soluble S^{2-} is oxidized and precipitated as S^0 in biofilms at some point where the O_2 reaches some (as yet unknown) threshold. S biofouling (white slime) occurs when Fe is typically absent or present in very low levels. Otherwise the S^{2-} is taken up as FeS.

The Redox Fringe

Redox zonation is a feature of aquifers, both those in close contact with the surface and those without such oxygen influence. Boundaries may be abrupt, especially in aquifers containing high levels of reduced organic compounds (Fish, 1993; Davis et al., 1993).

The boundary zone between zones containing and depleted of free dissolved oxygen appears to be an important environment for the mass occurrence of FeIII-precipitating biofouling organisms. This oxygen-depletion boundary in groundwater is referred to as the "redox fringe" by Cullimore (1993) and is the region in which dissolved Fe(II) species are oxidized to Fe(III). Of course, the precise Eh values in which Fe, Mn, and S transformations occur also depend upon pH and other physical parameters (Hem, 1985). Valid Eh determinations are notoriously difficult to make in natural and contaminated groundwater due to the variety of redox-active compounds present (Fish, 1993).

Hydrolysis of Fe(III) leads to formation of complexes of Fe(III) oxides with bacterial ECP. Mn(IV)-oxide formation, complexation, and precipitation occur in much the same way. Microbially facilitated Fe(II), Mn(II), and S^{2-} oxidation occurs, followed by deposition of Fe(III)- and Mn(IV)-oxides and elemental S. Figure 9 is a schematic of redox zonation around a pumping well.

An example of an apparent engineered redox fringe is that induced by VYREDOX Method installations. Hallberg and Martinell (1976) described the basic function of the VYREDOX Method, which involves injecting air into the aquifer around a production well. This effectively spreads out the redox fringe, inducing Fe and Mn oxidation and precipitation in the aquifer media away from the production wells. The aquifer then acts as a large sand filter to which the Fe(III) an Mn(IV) precipitates adhere. Gottfreund, Gottfreund, and Schweisfurth (1985) conducted studies in Germany that seemed to confirm a microbial component in Fe and Mn precipitation in the aquifer around VYREDOX-equipped wells. Information on S precipitation in aquifers is much more ambivalent (Fredrickson et al., 1989), but is commonly observed in wells.

Effects on Performance of Well Systems: A Summary

The extent and effect of Fe, Mn, and/or S oxidation and precipitation and associated biofouling in any particular situation depend on a variety of environmental, hydraulic, and use factors in the well or the downstream receiver of its production, such as a treatment system. These effects may be dramatic or hardly noticeable in the short term.

Probably the most adverse effect of Fe and (theoretically) Mn oxidation and biofouling is systemic plugging through an entire aquifer. Previous studies with core samples have shown that significant plugging, reducing the permeability of sediment cores by 65 to 99%, may locally occur (Characklis et al., 1987). However, in the limited studies available, Fe and Mn biofouling, like S biofouling, is rare in aquifer samples away from the near vicinity of wells unless oxidation is deliberately induced (e.g., VYREDOX installations or aeration for *in situ* bioremediation). However, widespread sediment clogging is certainly possible based on the example of massive, low-permeability Fe(III)-oxide ores deposited in sandstones and formerly unconsolidated deposits that typically serve as aquifers. Such ore deposition is considered to have a microbiological origin (e.g., Lovley, 1991).

Processes occurring in VYREDOX-treated unconsolidated aquifers are also likely to occur in aquifers aerated for *in situ* bioremediation, and probably in any drawdown fringe. Likewise, the Eh-pH and filtration conditions of a biological iron filter plant, such as described by Mouchet (1992), are favorably recreated in the redox fringe around filter-packed wells and subsequently in downstream treatment systems — this without the efficient backwash systems designed in engineered filters to keep the filter media clear.

The more typical case is that aquifers with waters that exhibit relatively low Eh levels overall contain an indigenous microbial community capable of reducing Fe(III), SO_4, and possibly Mn(IV). The Fe-, Mn-, and S-precipitating biofouling itself, associated with aerobic metabolism, occurs at the oxidation-reduction interface (redox fringe) near the water table, wells and springs, and along organic contaminant plume fronts as described. In well systems, S slimes that impede flow seem to occur (1) when there is significant dissolved S^{2-} and O_2 available in the pumped groundwater and (2) in boreholes or downstream at meters, particulate filters, or other restrictions. Such well systems provide conditions similar to those at spring outfalls where sulfur slimes are found naturally.

In the well itself, biofouling phenomena, including associated Fe hydroxide precipitation, may encrust or loosely plug well borehole intake areas and screens, pumps, and other equipment. The initial process is the formation of a biofilm on surfaces in the well (casing, screen, pump) and the aquifer in the vicinity of the well. In a relatively oxygenated well environment, typical of pumping wells, Fe-, Mn-, and S-depositing biofilms may form within weeks. The time course of this process, resulting in water quality or pumping problems, may vary considerably.

Hydraulic Impacts

Hydrogeologic conditions affect the impact of Fe, Mn, and S biofouling on the well. For example, wells in highly transmissive gravels or rock aquifers do not tend to plug at the borehole wall even if they may be experiencing significant biofouling. Fractures and solution channels intercepted in these formations typically have such a large volume (and volume-to-surface ratio) that noticeable plugging is improbable in most cases. On the other hand, poorly or moderately transmissive porous media aquifers and filter packs have low volume-to-surface ratios and are more vulnerable to plugging. Zones in which Fe(III) oxyhydroxides are precipitated may be quite narrow, resulting in the potential for aquifer-clogging bands of Fe oxides.

Biofouling effects on pumps and discharge pipes, as well as downstream water collection and treatment system components, can be dramatic. Fe, Mn, S, and combination deposits break loose and enter the pump, leading to clogging problems. Biofilms on interior pipe walls become increasingly hard or thick over time. Corrosion tubercles or iron build-up at nodes on steel pipe increase hydraulic resistance by reducing diameter and by increasing roughness of the interior surface of the pipe, thereby dramatically increasing the energy cost to pump.

These effects may not be apparent until they are well advanced unless regular monitoring of system water quality and performance is carried out. Modern turbine pumps and other pump types, such as gas-driven models, used in monitoring sampling, are robust and may not exhibit the symptoms of deterioration for long periods.

Sample Quality in Monitoring Wells

Well deterioration can result in sample quality degradation in a number of ways:

1. Solid material entrained in sampled water can adsorb dissolved compounds or interfere with analytical quality.
2. Chemical encrustation reflects shifts in redox potential and pH (and sometimes pressure) that may alter the solubility of compounds in groundwater.
3. Pump and screen corrosion adds metals such as Ni and Cr to groundwater, where they might not occur otherwise. Corrosion also results in excessive metallic oxides in suspension.
4. Due to the reactivity of metallic oxides such as Fe hydroxides, their principal effect when present is to act as chemical sieves, adsorbing reactive inorganic species and organic compounds. Pumped samples from Fe-biofouled wells then do not provide analytical results representative of the aquifer beyond the well's area of influence.

HUMAN FACTORS

It has sometimes been puzzling why two similar wells, both with well deterioration problems (and especially in the case of biofouling development), experience very different symptoms. The explanation may be in well design, construction, and operation in many cases, combined with subtly different in-ground conditions, such as local differences in hydraulic conductivity. These may result due to formation conditions or development differences. Such design and operational aspects are primarily in the realm that human activity affects directly — that is, people can influence them. Toxicity and pathogenicity are also human concerns, both for the workers operating these systems and their effluents.

Management and Operational Overview

The design and construction of monitoring and recovery wells is not the primary focus of this book, but proper well design, construction, and development are important in prevention and planning for maintenance as described in Section II. These are areas in which engineering knowledge, skill, and application have an impact.

Good well design has several interlocking aspects. Wells designed to resist corrosion and permit reasonably free but laminar flow to the well usually provide water with the minimal possible drawdown and oxidation at the intake. These are less likely to plug quickly. Good screen and pack selection minimize potential sanding problems.

The construction process can defeat a good design if packs, screens, or casings are installed in a hasty fashion or development is inadequate. Adequate development may be neglected because it takes more time and skill than may be available. Project supervisors may fail to demand adequate development because of apparent time pressures or they want to avoid "disturbing" the monitored formation. If they do not have experience with the installation of pumping wells, field supervisors may not have a grasp of the time required to fully develop a well: the scale is hours, not minutes.

Well operation impacts biofouling in particular by providing conditions that enhance or discourage biofilm formation and build-up. Cyclical pumping or long periods of idleness (i.e., stagnation) promote biofilm formation and associated problems. Excessive pumping drawdown introduces more oxygen into the system, increasing the rate of FeII-FeIII oxidation (both abiotic and microbial). Such operating profiles are, of course, the operational descriptions of typical monitoring and recovery or plume control well systems.

System operational design is also a factor. In multi-well systems with automatic pump controls, stronger wells may produce more water to compensate for other wells with falling production. The result may be that strong pumping occurs around productive wells, while the plume quietly bypasses clogged wells. There has been virtually no research into these phenomena, so we just don't know for sure. However, active site management can make adjustments to compensate for these changes. Proactive management makes use of performance and water quality monitoring to detect such situations and prescribes a maintenance strategy to deal with them (Section II).

Health Concerns Relating to Biofouling

Traditionally, biofouling has been considered primarily an engineering or operational, and not a high-priority health concern. Likewise, bioremediation is encouraged in contaminated groundwater and soils, with only the use of genetically engineered microorganisms actively discouraged at present. Wherever mass growth of microorganisms is encouraged or tolerated, operators of groundwater remediation treatment systems need to consider the health aspects.

Pathogens

There is a tendency in bioremediation potential studies to pass over identification of microbial types as "too expensive." However, if potential pathogens are present, especially in concentrated slugs in pumped groundwater or aerosols from aeration towers, there is a potential hazard to personnel directly exposed. For example, Fe-precipitating and reducing bacteria frequently found in mass microbial development include genera that are better known as including potentially enteric and respiratory pathogenic types (e.g., *Escherichia*, *Clostridium*, *Klebsiella*, *Pseudomonas*, and *Serratia* spp.) (Mallard, 1981; Lovley, 1991).

Mixed microbial populations in which the components produce mutually beneficial products or conditions are common in the environment. Such mutualistic consortia may contribute to the survival of other virulent or opportunistic pathogens traced to well water supplies. For example, besides the Fe-manipulating potential pathogens just discussed, *Legionella pneumophila* (a respiratory pathogen) is assisted by association with other bacteria.

Legionella is common in soils and many other environments, including cooling tower systems (Stout, Yu, and Best, 1985) that are analogous to the commonly used VOC (volatile organic compound) stripping aeration towers on environmental sites. Other common sources of possible human contact are wet soils and poorly maintained water systems with dead ends and biofouling. *Legionella* bacteria may also be found encysted within their predators: larger bacteria and protozoa in soil and biofilms.

Enhancement of natural *L. pneumophila* populations in natural soils undergoing aerobic bioremediation would also be expected. Wet, organic-laden soils with high microbial densities are favorable habitats for their mass growth. The presence of such *Legionella* in airborne dust and aerosols around such remediation facilities has to be considered a potential inhalation hazard, primarily to people with impaired immune responses, including smokers, who typically lack nose and throat defenses against airborne bacterial infection.

Toxic Accumulation

Biofilms and their associated reactive metallic oxides can accumulate toxic metals and other toxics at levels higher than in the bulk groundwater itself. The reason may be for some nutrient bioaccumulation purpose or as an antipredatory effect (Cullimore, 1993), or simply due to the reactions with the charged surfaces of metallic oxides as previously described. The benefits of binding toxic U and Au to cell membranes and sheaths are not at all clear at this time.

There have been occasions where Fe-biofouled wells were found to have become point sources of heavy metals contamination through accumulation over time. These metallic elements can originate from weathered granitic or metamorphic rocks, ore bodies, and other sedimentary rocks derived from them (such as shales), or anthropogenic sources such as buried sources of metallic or radioactive waste.

Both the biofilm slug effect and microbial mobilization can result in intermittent high metal or radionuclide concentrations in the treated water. This can result in a declaration of nonperformance by the remediation site's regulatory agency that probably would not have occurred if the biofouling were not present.

Chlorination of Organic Chemicals

Chlorine and chlorine-based disinfectants can react with components of the biofilm or hydrocarbons in the groundwater to produce halogenated organic compounds. Chlorine substitutions of hydrogen make the organics more resistant to safe degradation and also more toxic. For this reason, chlorination is routinely excluded as a well treatment method for monitoring and extraction pumping wells in groundwater contamination control schemes.

ECONOMIC IMPACTS OF WELL DETERIORATION

Identifying Costs of Well Deterioration

There are both direct and indirect maintenance costs associated with clogging, sanding, biofouling, and corrosion. The direct impacts include reduced performance of the well and well equipment such as pumps through clogging and corrosion. Indirect impacts include increased costs of operation through reduced efficiency, increased capital and operational costs due to corrosion, clogging and/ or encrustation, coating and fouling of resins, shortened filter cycles, and additional chemical costs. The most costly result is of course the failure of the system to perform its design function, such as adequate monitoring of indicators of groundwater quality, plume control, or remediation.

Types and Dimensions of Costs

A large percentage of the direct costs of well deterioration is due to side effects of Fe, S, and/or Mn biofouling, such as corrosion of pumps, discharge pipe, and screens, and clogging of screens, pumps, and pipelines. Costs may take the form of well rehabilitation and pump and column pipe repair or the abandonment of old wells and drilling new wells. Pumps in particular are expensive to fix or replace.

At this point in the damage-assessment and decision-making process, a decision may be made on actions to deal with the deterioration. Will the well or system be operated in its crippled state? Or will it be decommissioned and replaced, or rehabilitated? Decommissioning as an option may be chosen if the system is deemed uneconomical (or otherwise not feasible) to rehabilitate. Rehabilitation may be chosen if it is technically feasible and desirable.

Rehabilitation is usually attempted for deteriorated wells once a performance problem is recognized. Usually, rehabilitation takes the form of cleaning and refurbishing pumps, replacing corroded pump discharge pipes and other appurtenances, cleaning out pipelines, and chemically/mechanically cleaning the clogging and encrusting material from the well. In the case of biofouling, rehabilitation is often only partially successful. Relief may occasionally last for years, but more typically a few months on environmental projects.

Rehabilitation of high-capacity public water supply wells, for example, is relatively costly: typically $3500 to over $5000 to refurbish a pump and $4500 to over $20,000 to completely rehabilitate a high-capacity well. The cost is similar

($8000 to $20,000 per well, regardless of the yield value) for hazardous and toxic waste recovery projects due to the usually advanced state of fouling in such wells, and the extra time and care involved in the cleaning process and control of fluids. Purge fluids often must be handled as liquid hazardous waste and may not be suitable for treatment with the on-site process. Workers may be required to work in protective gear, and there is of course the time cost of suiting up, decontamination of people and tools, and the other well-known rituals of contamination containment. The alternative, however, is an even costlier compromise or loss of well system performance or degraded water quality.

An estimate of the costs of neglecting to detect and control well deterioration is warranted, even though current information is sparse. Based on a variety of available technical and economic data, Smith (1990) estimated that the direct cost of well deterioration for U.S. water supply utilities and irrigators to be conservatively in the range of $200–285 million annually and close to $1000 million when private water supply wells were included. The total economic cost to environmental control projects is not yet calculated, but easily could approach these figures.

Beyond the cost of replacement and rehabilitation is the cost in terms of reduced performance. For comparison, Howsam and Tyrrel (1990) estimate that 40% of water supply wells worldwide are operating inefficiently or are out of commission due to well deterioration. Contractors performing environmental well restorations consider that the large majority of environmental pumping wells are operating inefficiently.

Efficiency is a large factor in pumping and other well operating costs. Helweg, Scott, and Scalmanini (1983) have provided an empirical formula for calculating operating costs. In an example of the use of this equation provided by Borch, Smith, and Noble (1993), a 250-gpm well pump operating at 60% efficiency would cost $10.00 per day to run. Increasing drawdown (s) from 50 to 75 ft would increase the cost to $10.94. Reducing efficiency to 25% (not atypical) and maintaining s = 50 ft (typical of pump and discharge pipe corrosion) increases the cost to $23.99 per day, all other parameters being the same.

Direct economic impacts are more difficult to judge for monitoring wells. As discussed previously, monitoring wells are widely assumed to provide water quality altered by the well environment (e.g., biochemical filters). The costs come in the form of the consequences of having monitoring points providing erroneous information: (1) questions about the validity of data in public hearings, legal proceedings, or regulatory actions, and subsequent costs of confirming data, or providing new monitoring points; (2) worse, actual arrival of a plume at a water supply well or ecological treasure because monitoring wells failed to provide the necessary early warning.

It might be worthwhile to mention that developing micropurging protocols does not seem to take near-well chemical alteration into consideration and may delay identification of monitoring well deterioration.

For recovery and remediation wells, reductions in efficiency are relatively rapid and steep. Recurrence of renewed deterioration after rehabilitation is also rapid since well cleaning is rarely very effective in removing clogging precipitates. This situation puts managers of remediation wells in the position of dealing with rapid and repeated declines, with well rehabilitation in each case being relatively expensive.

The Helweg, Scott, and Scalmanini (1983) operating costs equation and other equations and nomagraphs like it are relatively simple and do not address additional "downstream" factors. Examples of these include: (1) increased chemical costs due to removing precipitated Fe or Mn; (2) labor and equipment costs involved in renewing biofouled water treatment systems, water treatment equipment, or encrusted pipelines; (3) excess consultant or contractor time devoted to fixing problems; or (4) business losses.

Business losses to remediation or environmental engineering companies may result from reduced client or regulator confidence. Business losses for facilities may occur if regulatory officials order a shutdown if groundwater contamination is not contained.

These factors have to be calculated on a site-by-site basis, using normal accounting methods.

Among the important cost considerations, human time stands above and beyond anything else. The costs of well materials, pumps, chemicals, etc. are relatively minor by comparison. For example, let's consider a fairly simple example. The reader can plug in relevant costs in their own situations and extrapolate outward for complex systems.

A Costly Example

Problem: A set of four remediation wells feeding a filter and stripper unit loses performance and clogs the treatment system and piping with iron precipitation. Rehabilitation becomes necessary as abandonment is not possible.

Time (never mind related direct and indirect costs):

1. A week for two people (site personnel) to take down and clean the filter/stripper system (clogged and coated by iron precipitation), with supervision and health monitoring.
2. 2 days to clean four wells plus 1 day each mobilization and demobilization (contractor).
3. Consultant time to draw up remediation plan for state approval (2 days) and to confirm system and filter performance (2 days).
4. Approval process triggers an OSHA site inspection, resulting in 2 days of managerial time, plus citations that have to be addressed.

Chemical/material costs:

1. Discard and replace filter media and aeration tower packing (secure disposal due to chemical residues).
2. Discard and replace accessible piping (secure disposal due to chemical residues).

3. Mechanical/acid cleaning of supply lines from wells (20 l of acetic acid), line brush, and rods or cables (clean and test for residues).
4. Containerize and discard acid sludge waste from lines.
5. Mechanical/acid cleaning of wells, 1000 l of acetic acid solution, including use of a jetting rig (rig and lines to be decontaminated after well cleaning).
6. Containment and secure disposal of 3000 l of purged pH 6.5 but turbid water containing the VOC contaminant.

The goal of Sections II and III is to introduce means of limiting these considerable and oftentimes unplanned operational costs. These sections include methods to compare maintenance costs vs. the costs of well deterioration.

II. Prevention and Maintenance

Poor performance and failure of monitoring and remediation well systems can be controlled or prevented. This can be accomplished via prevention in design and construction, a recognition of well deterioration factors, monitoring for problems, and preventive actions when problems are detected. If well systems do deteriorate, a limited number of rehabilitative options are available (Section III). Monitoring and preventive design and maintenance are preferable on both cost and operational bases.

3 The Basis for Controlling Well Fouling under "Environmental" Conditions

This chapter will consider prevention and control of well-deteriorating conditions, including detailed information on preventive maintenance monitoring. Rehabilitative strategies are discussed after that. If you skip ahead to "cures," come back to learn about ways to avoid a crisis the next time.

There is an extensive literature and body of unpublished and uncelebrated practical experience with prevention and removal of clogging conditions in water supply wells. There is a large parallel body of experience in corrosion prevention in oil field and marine engineering systems. These experiences are directly transferable to the design, operation, and maintenance of monitoring and pumping wells on environmental sites, with some exceptions. However, the quality of existing information varies greatly, and site consultants or managers should not proceed based on armchair research alone. They should also take into account the experience now being gained by the handful of companies that are actively performing environmental well rehabilitation and maintenance. Based on this experience, there are several ground rules of well maintenance and rehabilitation for environmental well systems:

1. The "best defense" is knowing about and acknowledging the presence of factors in the well and aquifer system that will cause clogging and corrosion. Knowledge can lead to either action or despair, but preferably preventive action to the degree possible.
2. Prevention, rehabilitation, and maintenance possibilities are all more limited than those possible for water supply wells and certainly those used in oil and gas production and reservoir flooding wells. Prevention is limited to a degree by the need to construct wells in sediment- and microbe-rich zones with a robust biogeochemical variety (see Chapter 2). Such situations can sometimes be avoided in water well design, whereas they can't always be avoided in environmental well planning. However, preventive well design is certainly possible also for environmental wells.

3. Maintenance is one area where much is possible, but it requires a commitment to doing what needs to be done. Maintenance is best implemented from the beginning, but can be implemented after deteriorated wells have been rehabilitated to slow or prevent recurrence of the problem.
4. Rehabilitation itself should be the last phase and last resort (before decommissioning or reconstruction) in the life cycle of a well system (Figure 10). It is never a permanent solution and has to be followed up by maintenance to be effective. Rehabilitation is frequently limited by environmental protection and safety factors when chemicals are present in the groundwater. It is also limited by the typically small size and relative delicacy of the wells.
5. Preventive maintenance and rehabilitation strategies are site specific and require fine-tuning as operators gain experience with deteriorating conditions on-site.

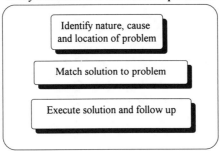

The purpose of maintenance or rehabilitation can be one or more of the following:

- ☑ *Process* control
- ☑ *Condition* management
- ☑ *Performance* recovery

Maintenance is intended to control *processes* that may cause undesirable *conditions* that can result in *performance* deterioration.

Rehabilitation is intended to correct an undesirable *condition* and attempt to recover *performance* after it has deteriorated.

In any maintenance or rehabilitation procedure:

Identify nature, cause and location of problem

Match solution to problem

Execute solution and follow up

Examples

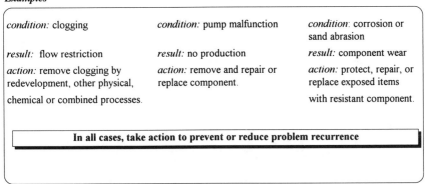

condition: clogging	*condition:* pump malfunction	*condition:* corrosion or sand abrasion
result: flow restriction	*result:* no production	*result:* component wear
action: remove clogging by redevelopment, other physical, chemical or combined processes.	*action:* remove and repair or replace component.	*action:* protect, repair, or replace exposed items with resistant component.

In all cases, take action to prevent or reduce problem recurrence

Figure 10.   Purpose of well maintenance and rehabilitation. (Derived from CIRIA Report, *Monitoring, Maintenance and Rehabilitation of Water Supply Boreholes*, © CIRIA 1994; Howsam et al., 1994.)

4   Prevention: Design and Construction Considerations

Prevention is the fundamental step in avoiding costly and project-threatening well deterioration and plugging. They may not totally cancel future well rehabilitation dates, but preventive strategies prolong well operational life and make maintenance and ultimately rehabilitation more likely to succeed.

The best protection against deterioration in wells and water systems of any kind is prevention. Prevention involves a combination of good design and construction practices, followed by preventive maintenance monitoring and treatment. Practicing prevention requires a team effort among well managers and operators, drillers, equipment suppliers, and consultants.

Prevention needs to start at the very beginning. Unfortunately, the usual situation at the present time is that the site operators are considering improvements in well performance after wells have deteriorated in performance or water quality. Even when reacting too late, it is useful to review good design and construction practices to assess (1) what went wrong or (2) how it can be done better the next time. Some problems are preventable from the beginning. Table 2 provides a review.

A crucial part of prevention planning is proper well design and construction. There has always been a temptation on many projects to attempt to save money on a monitoring or remediation project by asking too much of very limited wells, using standard pack and screen sizes, inexpensive pumps, and shaving on development time.

Monitoring and recovery/control wells are no different than water wells in that quality well design and construction contribute to long and trouble-free service. For monitoring wells, quality is absolutely essential because so few effective options exist for rehabilitation.

Remediation well systems bear close resemblance to geotechnical dewatering systems in design, operation, and the environments to which they are exposed. Powers (1992) ably discusses design and operational aspects of these systems.

Two useful reference standards in construction of monitoring wells are ASTM D 5092-90, *Standard Practice for Design and Installation of Ground Water Monitoring Wells in Aquifers* and D 5521-93 *Standard Guide for Development of Ground-Water Monitoring Wells in Granular Aquifers*. These standards

Table 2   Preventable Causes for Poor Performance

1. Poor selection of well/wellfield location or selected aquifer intervals (not often avoidable in environmental projects).
2. Poor design of intake area (allows silt infiltration or is too restrictive, adding to clogging).
3. Inappropriate selection of materials: corrosion or collapse.
4. Mistakes in construction.
5. Lack of sufficient well development.
6. Inappropriate pump selection.

reflect modern practice in monitoring well construction and development, and many features can also be transferred to pumping wells.

These standards (and the ancillary ASTM standards available and under development to support them) are not intended to be encyclopedic, or even always highly specific. Design, construction, and development of course should be based on site-specific information as the standards recommend. For example, even when casings, pumps, and gravel packs have been lovingly selected for material compatibility and corrosion resistance based on the literature and developing standards (using good water quality data), poor performance may still occur. If biofouling is going to coat everything (e.g., blanking off a reactive casing or, alternatively, accelerating corrosion), this should be known up front.

PREVENTION IN WELL DESIGN AND CONSTRUCTION

The purpose of water well construction is to provide a water supply well that will efficiently provide good quality water over a long service life — on the order of decades. Monitoring wells, like water wells, are usually intended to have a long service life — on the order of decades. Trends in groundwater quality at a location, such as rates of plume degradation or increases in saline intrusion, can often only be ascertained based on long study of data from consistent monitoring points. As has often been pointed out, however, monitoring wells are not usually designed to provide the maximum water production.

Recovery and plume control wells, while both extraction pumping wells, have divergent lifestyles. Recovery wells, for example, are optimistically expected to have short design lives. However, current experience indicates that both recover/remediation and plume control are long-term processes for most sites. In any case, they must be as productive as the formation permits and do their job as long as needed.

Another difference from water supply wells for all classes of environmental wells is that they have to be in specific places for the job requirement. This particular zone or aquifer has to be sampled or pumped down regardless of the long-term impact on the well.

Quality well design and construction assists in the prevention of encrustation and corrosion problems during the life of wells. Using quality materials, using care in construction, and proper sealing and disinfection, the designer and driller can help to assure a long and less-troublesome life for the well.

Well Design

Proper well design, in addition to determining the depth and diameter for the purpose of the well, includes casing selection, appropriate intake section design, grouting to prevent infiltration around the casing and between sampled intervals, and procedures for well development, testing, and removal of other introduced materials (drilling fluid, surface soil, etc.). Unlike water wells, disinfection is not likely to be a necessary step. All of these tasks are necessary to achieve the optimal performance. Neglecting any of them is false economy.

While monitoring and many recovery/control wells are often not great producers or efficient in a water well sense, within the constraints imposed by the situation (such as the need to sample specific intervals with minimum oxidation), they should still be designed with optimal efficiency in mind. The main reason is that designing for maximum efficiency helps to minimize biofouling and encrustation, as well as oxidation, effects to the extent possible.

In monitoring well doctrine, materials selection is made so that materials do not interfere with analyses. A plus from a maintenance standpoint is that noncorrodible and nondegradable materials are also those that provide resistance to biofouling and corrosion.

Table 3 summarizes general well design and placement guidelines. This advice is echoed elsewhere (e.g., ASTM D5092-90; Powers, 1992).

Casing for Well Completion

Casing is used in wells to (1) provide a stable hole and (2) seal the walls of the hole to exclude undesirable water. In a pumped well, such as a recovery well, the casing must also house and protect the pumping equipment.

Nielsen and Schalla (1991) provide an in-depth discussion of casing selection for monitoring well design. The casing in a monitoring and/or pumped extraction well must have (1) a sufficient diameter to accommodate pumping equipment and instruments, (2) strength to withstand forces during emplacement and use, and (3) ability to resist corrosion, heat, abrasion, and other causes of well deterioration that effect sample quality. In monitoring wells, casing diameter is often restricted to limit the amount of well purging necessary and in some cases to minimize the drill cuttings that have to be drummed and landfilled during well construction.

From a maintenance standpoint, diameter, strength, and corrosion resistance are important as well. Casings that have sufficient tensile and compressive strength and corrosion resistance are unlikely to fail catastrophically. Diameter is a factor in limiting which redevelopment tools may be used, as well as determining the volume of purge water that may need to be handled after a treatment. Casing material should be selected on a site-specific basis, taking into consideration water quality and hydrogeologic conditions.

Table 3   Well Design Guidelines

1. Well design should be done by qualified groundwater professionals (engineers or hydrogeologists experienced with well hydraulics and construction), basing the design on the needs of the project and available industry standard practices (e.g., ASTM).
2. Perform chemical and microbiological analyses of the water to determine the characteristics of the water in the aquifer relevant to well design and performance. Repeat once operations begin to make any adjustments. If not done, unanticipated problems follow.
3. Good material selection will provide good service, with installation price as a secondary consideration. The motto is "penny wise, pound foolish" in design and materials. It is important to closely match the material selection to the probable effects of the constituents in the groundwater of the site in question.
4. Select and install pumps in pumping wells designed for the pumping rates and environment planned. Material selection concerns for well construction apply to pumps. Choose pumps designed with the necessary reliability and durability for these tough environments.

All water-well grade casing thermoplastics have almost total corrosion resistance. However, they may be subject to attack and softening by certain organic solvents, just as if they were being solvent-joined. Tetrahydrofuran, methyl ethyl ketone, methyl isobutyl ketone, and cyclohexane, for example, if present in parts per thousand or percentage concentrations, may solvate thermoplastics (Nielsen and Schalla, 1991). On the other hand, they may be highly resistant to other nonpolar solvents such as gasoline components. In the typical parts per million and billion ranges of concentration, such casing degradation has not been reported. On the other hand, dissolution of cement bonds has been reported, so solvent-joining should be avoided in lieu of threaded joints.

Heat is a consideration for wells on projects using heat-amended remediation or requiring cement grouting. Plastics chosen (if stainless steel is not an option) may have to be thermally stable: not likely to become reactive or to physically deform. PVC is not thermally stable, for example, above certain temperatures, depending on the composition. Thermal stability data provided by a manufacturer should be for the expected lifespan of the well. Short-term temperature resistance data is not relevant for long-term exposure.

Stainless steels have good corrosion resistance in general, but most especially if oxygen is present. The way that stainless steel corrosion resistance works is that a layer of metallic oxide is deposited on the metal surface. Under reducing conditions however, which prevail as biodegradation of organics occurs, stainless steel corrosion resistance may be impaired. For this reason, steels of all sorts (including stainless) should be considered susceptible to corrosion under groundwater conditions in which organics are present. In a case where Type 304 may be specified under uncontaminated conditions, a higher grade such as Type 316 may be needed.

Alternatively, plastics, at sometimes a third or less of the cost (only in materials), should be used if groundwater quality conditions will promote corrosion. However, strength and thermal considerations may preclude the use of available thermoplastics. Thermal and collapse resistance depend on the plastic pipe material used. As with steel casing, the diameter and wall thickness of the casing pipe determine the hydraulic collapse resistance of plastic pipe. Tables for various plastic and steel casings are provided in Driscoll (1986) and Roscoe Moss Co. (1990). Standards and data are also provided in the relevant ASTM pipe standards (American Society for Testing and Materials, Philadelphia, PA).

As for plastic casings, threaded joints are preferred for stainless steel casing since welds are more susceptible to corrosion. Field experience shows that stainless corrosion begins in heat-affected weld regions.

Well Hydraulics and Efficiency

Good hydraulic flow characteristics at the well contribute to good efficiency and reduced maintenance problems. In particular, for wells that may experience biofouling, good hydraulic efficiency reduces the impact of clogging and allows time to begin a rehabilitation program.

It is beyond the scope of this book to discuss well hydraulics in detail, and it is well covered elsewhere. However, efficient screen selection, reasonable pumping rates, and thorough well development contribute to optimal hydraulic efficiency, which in turn improves the reliability of monitoring because a good formation-well contact is established. Good well hydraulic characteristics depend on some basic understandings of the aquifer.

Well Screens and Intakes

Wells generally can be divided into two categories based on the intake type: screened or unscreened. Aquifer formations that require screens are usually those that are unstable and must be held back from the borehole. These include sands and gravels as well as unstable and weathered rock formations. This is the most typical aquifer tapped by a monitoring or recovery well. Filter-packed screens are considered to be assumed in monitoring well design (cf. D5092-90). Recommendations for good monitoring well screen and pack design has been ably described elsewhere, notably by Rich and Beck (1990), Schalla and Walters (1990), and Nielsen and Schalla (1991).

Screen Design

The ASTM monitoring well design and construction standard practice (D 5092) currently defines what is good screen and pack design and selection for monitoring wells. It is intended that in the future, further ASTM standards under development will support the core D5092 standard. However, even when these standards appear, it must always be realized that well design has to be site specific in nature.

Screened wells in environmental applications also can be divided into those that are (1) always submerged and (2) sometimes exposed to unsaturated conditions, such as product recovery wells. In standard water well design doctrine, dewatering of the screen is to be avoided. This ideal is often not possible in dewatering wells and product recovery wells. Dewatering wells constructed over impermeable sediments or rock may have to draw down completely to produce the desired water level. Recovery wells also often have to provide contact with the top of the local water table to permit skimming of light hydrocarbons.

In these cases, engineering involves "designing for failure" more than in any other case. Designers must realize that these wells will biofoul and plug over time, design them accordingly, and make plans from the beginning to perform regular maintenance.

Within the constraints of the well purpose, "designing for failure" can take the form of expanded filter pack and associated larger borehole diameter to permit more complete rehabilitation. Screen materials should be chosen to resist the expected rehabilitation actions. Screen inflow modification may be employed to slow velocity and clogging (see Pump Protection, this chapter). Well termini and hookups should be designed to permit access for rehabilitation.

Quality manufactured water well screens can be relatively expensive. Regardless, they are preferred for long life since they have good hydraulic efficiency. Stainless steel, plastic, and even fiberglass models resist erosion and corrosion much better than slotted steel pipe or galvanized wire screens.

In any case, the slot size should be small enough to contain most filter pack particles, but not too small. The slots should be uniform in width and free of shavings or weld spatter. Various text sources show examples of screen designs (e.g., Borch, Smith, and Noble, 1993).

Slotted screens should be installed with a sand filter pack per ASTM D 5092, while wedge-wire screens may also be naturally developed (Driscoll, 1986). Screens may be installed with the casing in rotary- or auger-drilled holes, or slipped inside cable tool-driven casings using the telescoping method.

Material Selection

With their higher surface areas, screen material selection is more critical than that for casing material. Stainless steel screens are often the material of choice for monitoring and especially extraction wells due to the ability to very precisely define slot size, the strength of even wire-wound screens to resist development forces, and their long-term durability.

PVC or fiberglass screens made from water well casing tubing are often superior to metals under chemically reducing redox conditions in which steel corrosion is accelerated. On the other hand, some solvents present may attack these materials. Also, deeper and hotter conditions restrict plastic and fiberglass use.

In the cases where plastic screens (thermo or composite) are unsuitable due to constituents, heat, or strength factors, stainless steel should be used. However, where reduction is possible (in the presence of hydrocarbons, especially), corrosion has to then be considered in maintenance planning.

One lamentable situation at present is the highly variable quality of fiberglass casing and screen products. In many cases, fiberglass composites have very favorable characteristics, with strength and heat resistance superior to comparable thermoplastics. However, the finish of produced items leaving exposed, friable glass fibers makes them unusable for monitoring and extraction purposes, as well as water supply production. If fiberglass is an optimal option, the quality of the available products has to be determined before specifying.

The use of mill or field slotted and perforated steel casing should be avoided in all cases, as should mild- or galvanized-steel wire-wound screens, even in short-lived product recovery wells, the reason being that biofouling and biodeterioration by-products can corrode these screens well within even the optimistic estimated short performance lives. Factory-punched louvered and wire-wound screens finished and coated to minimize corrosion attack buy time against the still-likely corrosion attack.

In considering projected lifespans of wells, consider: just when has a pump-and-treat or *in situ* remediation project been completed on the predetermined schedule? What happens when a 2-month job extends to a year and you have a screen with a 2-month lifespan? The costs of restoring such a well or dealing with corrosion products is certainly much higher than simply specifying corrosion-resistant materials in the first place.

The sand or gravel used in filter packs should be clean — free from organic soil that may cause clogging and feed bacterial growth (the native soil will be lively enough). The filter material should be as uniform as possible, with a uniformity coefficient of 2.5 or less, preferably 1.0 (ASTM D 5092). Particle size should be selected to filter the formation without silt-packing at the borehole-pack interface. ASTM D 5092 specifies that the pack material should be selected based on sieve analysis to pass the 30% fraction (retaining 70%) and the screen should retain 90 to 99% of the pack. ASTM standards in development should further refine these recommendations.

Unscreened, open-borehole completions should include a casing or liner to the pumping water level at the least, unless low specific gravity product has to be recovered at the bedrock surface or interface. Small water-bearing fractures or screened zones above the selected pumping zone should be monitored separately and not allowed to cascade. Besides adding to water to be treated or interfering with sample quality, the water cascading down from these small seeps encourages bacterial growth and fouling.

Grouting and Well Sealing

In addition to basic quality well construction considerations, grouting helps by preventing the flow of bacteria-laden shallow water down into the aquifer and intake area of the well. It also helps by limiting contact of such water with the metal well casing, and thus limiting corrosion.

Again, this is not essentially a manual on well sealing, but several considerations should be mentioned. Grouts should be (Gaber and Fisher, 1988):

1. Of low permeability (lower that the surrounding earth materials).
2. Capable of bonding well with *both* the casing and earth materials.
3. Capable of setting up to strength quickly to permit well completion (including development) without excessive delay.
4. Chemically nonreactive with formation materials, constituents, and well materials.
5. Easily mixed and pumped in a reasonable time into the annulus.
6. Unlikely to penetrate far into permeable formations.
7. Easily cleaned and safe to handle.

The reader should stay current with developments in optimal grouting (e.g., Edil et al., 1992; and Lutenegger and DeGroot, 1994).

Both cement and bentonite grout mixtures should be properly mixed and placed in the well so that the space between the borehole and the casing is completely sealed per evolving recommendations and standards. (ASTM D 5092-90 is the most recent applicable standard for well construction grouting, but it is not the final word.) Unused boreholes and wells should be promptly and properly decommissioned to control inflow of oxygen and nutrients to the aquifer, which may affect fouling at other wells.

Well Development

Well development is the action of removing drilling damage and additives such as drilling mud from the intake area of the well and the surrounding aquifer. By doing so, development helps to restore the physical characteristics of the aquifer to the pre-drilling situation, provide a good hydraulic contact with the formation, and to remove fine cutting and formation material from the well vicinity.

In addition, development action in natural development situations sets up a gradation of particle sizes that tends to keep fines away from the well screen and improves permeability. Development in filter-packed wells helps to set up this gradation at the interface between the pack and the formation. Redevelopment, the process of applying physical development methods reactively, utilizes the same principles as those for initial well development.

Reasons for Development

The objective of well development for monitoring wells is to improve the ability of the well to provide representative, unbiased chemical and hydraulic data. It does this by helping to provide a suitable hydraulic connection between the wellbore and the surrounding formation so that natural groundwater can flow to the well, providing more accurate sample quality (Macaulay, 1988, D 5521).

Proper well development becomes more critical for monitoring wells in other ways. It minimizes the potential for damage to pumps, samplers, and sensors from abrasive particles. Those used in monitoring are often less resistant to damage than are water supply pumps. It also helps to at least initially minimize biofouling effects by removing bacteria-laden drilling mud and make-up water used in drilling, as well as contaminants such as compressor lubricating oil. By opening up the aquifer, development also helps to limit or delay the clogging impact of biofouling when it does occur.

Well development methods are well described in other literature, including Driscoll (1986), Roscoe Moss Company (1990), and Borch, Smith, and Noble (1993). These descriptions are oriented toward water wells, but most of the principles are the same as for any type of well expected to produce some fluid (Craig, 1991). Some discussion of development of monitoring wells has taken place in the published literature (e.g., Kill, 1990; Craig, 1991). ASTM Standard D 5521-93 provides standard guidance for monitoring wells in granular aquifers (a standard guidance for "karst" aquifers is under development). No such standard guidance should be considered limiting for methods used in environmental pumping wells. D 5521's intent is provide a minimum standard guidance for development, not to constrain methods that can be used.

Development Methods

There are a variety of development methods in use. In each case, the development of a fluid velocity in the near vicinity of the borehole is involved. Water is propelled out of the wellbore and flows back in, breaking up films and mixing up the aquifer material and filter pack. There are many variations on basic approaches, and some methods are more effective than others. Preferably (and essentially in more delicate wells), the process starts gently, increases gradually in intensity, and continues as long as improvement results.

One bit of information to remember is that proper well development for pumping wells takes time. It usually cannot be accomplished in half an hour; it usually takes some hours to fully break up drilling damage and cause the desired sorting of formation particles. Another factor to remember is that not all formations can be satisfactorily developed. Wells in intervals consisting mostly of very fine particles may never develop properly, and further development may just increase turbidity.

In monitoring and remediation well situations, several methods are available. Overpumping (which includes bailing), the most common form of monitoring well development, brings material into the well for removal as water is pulled into the well. A disadvantage is that overpumping and bailing tend to pack formation fines against the filter pack.

Conversely, back-flushing packs fines against the borehole wall. Bailing and air surging without pumping are slow and do not permit good feedback on events down hole. Air surging alone can also pack fines against the borehole wall and otherwise cause more damage than it solves in relatively delicate monitoring well environments.

For typical monitoring and recovery well situations, two methods probably stand out as the most practical and safe for development: (1) jetting with airlift pumping and (2) double surge with airlift or eductor pumping. Both provide the in-and-out motion necessary to properly develop wells. When properly used, both provide sufficient agitation to clear fines from the formation material. Pumping clears fines from the well.

Jetting: Jetting alone without pumping will agitate formation material and dislodge fines, but tends to pack debris against the borehole wall and introduces chemically altered water to the formation. Simultaneous pumping, usually by airlift, alleviates this problem. When working together, the jetting and pumping sets up a circulation, with the jet pumping water under pressure into the formation material and the water returning due to the suction pumping action. This removes the foreign water and fines and drilling debris. Jetting requires a high-pressure water pump and air compressor for airlift (or an installed pump). A circulation can be set up once upward pumping is initiated. The jetting tool itself can be as simple as a sealed length of pipe with drilled holes of a proper diameter and orientation (for balance) to provide sufficient volume and velocity against the screen (Figure 11), and can be of any diameter. Small monitoring well diameters provide an engineering challenge, however.

Jetting is most effective in V-slot screens and less effective in machine-slotted or louvered screens due to jet deflection (Driscoll, 1986), advice repeated by Kill (1990) and supported experimentally by Fountain and Howsam (1990). Jetting is also really only effective in relatively permeable formation materials and filter packs due to the limited penetration of jetting flow into the formation material.

Any fluids introduced must be of known and acceptable quality and developed out as soon as possible. Air used for airlift must be filtered to remove any residual compressor oil. If introduction of any air or altered fluid is unacceptable, jetting is usually precluded as an option.

Surging: Formations monitored in environmental studies and clean-ups tend to contain a high percentage of fine material, and well screens may be correspondingly fine with very limited formation contact. Surging is more easily adapted for these low-permeability, high-percentage fines formations than jetting and does not require high-pressure pumping or the injection of foreign water. While jetting necessarily involves mechanical pumping, surging should also be done with power equipment, as hand surging is too difficult to sustain to be effective, even with young weight-lifting graduate students on hand.

The double surge-eductor pumping method helps to concentrate the surging action of the swab, and pumping brings loosened material out of the well instead of merely washing it back out on the downstroke. Tools for this purpose are manufactured by, for example, Aardvark (Nuckols, 1990) (Figure 12) for small-diameter (2½-in.) wells and could very well be adapted for less than 2-in. I.D. This

High velocity jetting tools

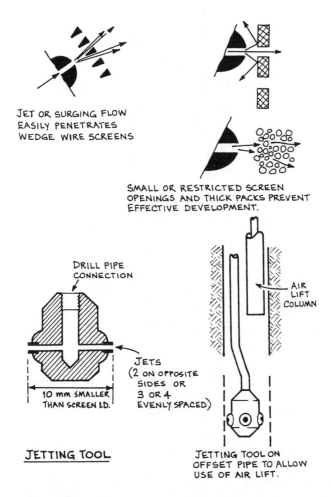

Figure 11.   A jetting tool. (*Australian Drilling Manual*, Eggington et al., 1992.)

tool consists of a dual wall pipe and double surge block. An eductor fitting is installed above the surge blocks in the pipe.

The manufacturer recommends that prior to the use of this tool (if physically possible), material inside the casing be vacuumed out using a suction tool. The development tool provides two actions: gentle surging provides the agitation to remove fines in the formation-screen-pack area, and a double surge block set-up concentrates the surging action. The velocity of air pumped down the outer pipe and past the mouth of the eductor sets up a vacuum in the surge zone that removes

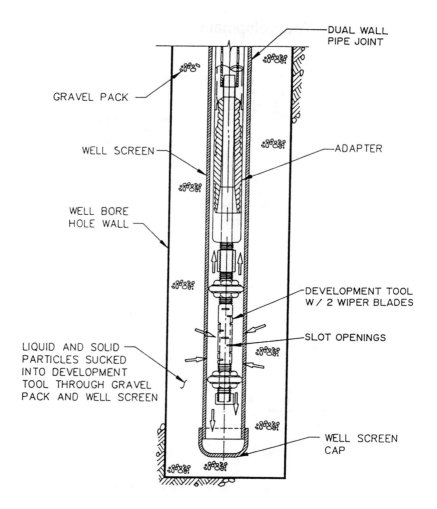

Figure 12.   A double-surge eductor tool. (Nuckols et al., 1990.)

water and solids (Figure 13). One caution about surging is that negative pressure
should not exceed the collapse strength of the weakest well structure component
(usually the screen or casing joints).

For both jetting and surging with pumping, solids can be removed from
pumped water via a sand cyclone (Figure 14) and development water analyzed to
monitor the well development progress. This is the only objective way to evaluate
well development effectiveness and decide when to stop (Eggington et al., 1992).
Also check other indicative water quality parameters such as pH, conductivity,
and Eh for stabilization.

Both surging and jetting have to be done with considerable care in typical
monitoring and recovery wells because they are not durably constructed (compared to municipal water wells), are possibly deteriorated due to environmental
attack, and may be more easily damaged.

▧ Air lift development

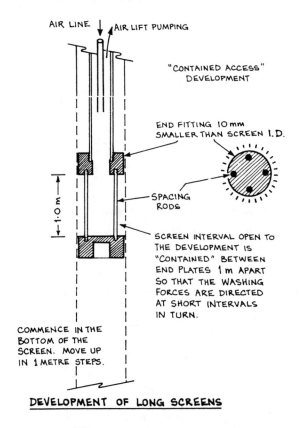

AIR LINE ↓ ↑AIR LIFT PUMPING

"CONTAINED ACCESS"
DEVELOPMENT

END FITTING 10mm
SMALLER THAN SCREEN I.D.

SPACING
RODS

SCREEN INTERVAL OPEN TO
THE DEVELOPMENT IS
"CONTAINED" BETWEEN
END PLATES 1 m APART
SO THAT THE WASHING
FORCES ARE DIRECTED
AT SHORT INTERVALS
IN TURN.

1.0 m

COMMENCE IN THE
BOTTOM OF THE
SCREEN. MOVE UP
IN 1 METRE STEPS.

DEVELOPMENT OF LONG SCREENS

Figure 13.   Surge-block eductor detail. (*Australian Drilling Manual***, Eggington et al., 1992.)**

Preventing Contamination During Drilling and Development

Drilling and development, as well as other well intrusions such as pump service, will never be sterile or really contamination-free. On the other hand, shallow aquifers have such large microbial populations that bacteria introduced during drilling are inconsequential anyway in the development of well biofouling. Still, steps are available to minimize contamination from tools and to limit drilling damage.

One favorite tactic in water well construction in preventing microbial contamination is the liberal use of chlorination in preparing tools, treating the well during construction, and in disinfecting gravel pack materials. Chlorination during monitoring and recovery well construction presents a host of problems, however, by drastically changing the aquifer environment locally and interfering with sample quality.

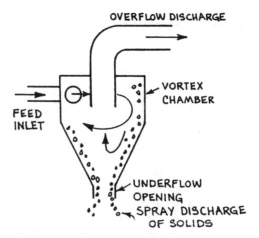

Figure 14.   Sand cyclone discharge for well development. (*Australian Drilling Manual*, Eggington et al., 1992.)

In any case, good cleanliness practice should prevail, as a matter of quality assurance. Tools, cables, pipes, wires, etc. should be free of visible dirt, oil, grease, etc. It is a good practice to keep tools up off the soil surface and to decontaminate tools before introduction to the well. The most effective decontamination for drilling tools is steam cleaning. While it is not reliable for sterilization, steam cleaning at least gets the tools clean and relatively free of organic matter. These steps are normally taken in any case on monitoring and remediation well installations. There is another reason to do so.

Development tools, cable tool bailers, and drill tools should be handled in such a way at the surface as to minimize contamination. All equipment and material to be installed in a well that is not prewrapped and assured clean (certainly a rare situation) must be decontaminated just prior to installation. This can be done by steam cleaning with an Alconox wash and filtered water rinse, such as described sketchily in ASTM D5088-90, followed by repeat steam cleaning to limit bacteria. However, it is important to realize that despite such measures (which will clean tools), there is no sterilization possible under any known procedure for drilling and well construction tools and equipment.

Not all monitoring wells can be installed with augers, so other more intrusive methods are employed. In spite of reservations about chlorination in environmental practice, prechlorinated water should be used for make-up water for cable tool drilling, circulation water for mud rotary, and in air injection. The water may be treated and stored in closed, disinfected tanks vented to allow chlorine dissipation. Air used in drilling should be filtered to remove compressor oil.

Simply keeping the solids contents of fluids to a minimum, minimizing the use of biodegradable polymers, and using mud tanks vs. open pits are good practices to limit contamination. Drilling additives suppliers have recommendations for mud control, tank designs, and chlorine levels. Extensive discussion of fluids control is provided in industry publications, most recently in Eggington et al. (1992).

Biodegradable polymer circulating fluids and lubricating oils have some following in the monitoring well drilling field, especially for coring, due to their capacity to break down, thus minimizing development and core interference. In the hands of experienced and skilled drillers, they have a place as long as it can be assured that the by-products are all removed. It should be remembered, however, that the breakers used also interfere with groundwater quality and that breakdown is seldom incomplete. Substrates remain that can be used by bacteria for food, and biofouling may be enhanced. Since biofouling interferes with sample quality, many of the advantages of polymers (relative to bentonite muds) in limiting formation interference may be negated.

For any well construction method, especially those using added fluids, proper well development to remove the "foreign" water is both feasible and desirable. Proper development is therefore, for many good reasons, a good place to start in the preventive maintenance treatment of a well.

PREVENTIVE PUMP CHOICES AND ACTIONS

Prevention should be considered in how the well is equipped and used after it is constructed. Preventive decision-making begins with deliberate decisions to choose minimal-maintenance pumps and to protect them to the extent possible.

Pump Selection

Pump selection can be something of an afterthought for remediation wells, although it is well thought out for monitoring in most cases. Pump selection options have improved in recent years with the introduction of improved designs based on years of trouble in the field.

The pump's service life is a definite consideration in well design and construction because of the high cost of pump repair and replacement, as well as the reliability of the installation. If a pump fails or works poorly, the well usually cannot do its job as a sampling point or means of managing a contamination plume.

In general terms, pumps should be selected for good service life under the conditions that will be encountered in the wells. They should also be protected as well as possible from unnecessary environmental attack.

For example, manufacturers and specifiers of well pumping equipment (e.g., Powers, 1992), as well as evolving consensus standards, recommend that wells be thoroughly developed to limit abrasives in the pumped water. Screens on the pump intake may stop larger particles that may come through. Certain pump types are designed to perform under sand-pumping conditions. There is certainly a benefit to this, in that it provides a margin for error. However, sand or silt pumping is an indication of a well structural problem that should not be ignored.

In extraction pumping wells in which silt intrusion cannot be fully controlled by the screen and pack, flow modification using suction flow control devices (SFCD) have shown good results in halting or reducing it to less than 1% of the previous level (Ehrhardt and Pelzer, 1992).

Most pump types do not do well when pumping dry, especially submersible centrifugal types. Output should be adjusted if necessary to avoid drawdown to the intake of the pump.

Power supply and electrical protection are important considerations in keeping pumps operating. Powers (1992) and Water Systems Council (1992) describe power supply and protection considerations in detail. Also, pump manufacturer sales and specification literature should be consulted for specific pump requirements and recommendations.

In any case, pump performance should be monitored and parts inspected on a regular basis. Table 4 summarizes possible pump problems. More on that is provided in the following and later sections on well maintenance and well reconstruction (Chapters 5, 7, 8, and 9).

Pumps in Monitoring Wells

Pumps used in monitoring may be either dedicated (installed permanently in a well and not moved) or portable (used repeatedly in different boreholes or wells). The choice is determined by the project requirements, and characteristics of these pumps are considered here from a maintenance point of view. The reader is recommended to industry literature and developing ASTM standard guides as they become available for pump selection from a sampling quality standpoint.

Dedicated monitoring well pumps: Monitoring wells that are designed for long-term use are frequently equipped with dedicated pumping-sampling installations. These are intended to work in-place for long periods, often in aggressive conditions, so their proper selection and maintenance is of interest for good service.

Most of these dedicated pumps are at present of the bladder-type design, with the pump at the designated pumping level, connected to the surface via discharge tubes and air lines used to transmit compressed gas to the bladder. Bladder pumps resist clogging, and the bladders generally are not subject to abrasive wear at the low flow rates used in groundwater sampling.

Stainless steel submersible centrifugal pumps designed for monitoring well applications (Grundfos Redi-Flo, Clovis, CA) are also showing good service in silty water, at least for short periods (Gäss et al., 1991). Manufacturers of neither dedicated bladder pump units nor submersibles report significant maintenance problems with the pump ends in general use if abrasives are limited. However, if abrasives are present, check valves and tubing may experience scour.

Most problems experienced with dedicated pumping units occur above the pumping water level. Both discharge tubing and air lines, as well as their fittings, are prone to freezing in cold conditions unless protected. Provision for heaters can

Table 4   Things That Go Wrong with Pumps

1. Poor selection of pump model for the working conditions (e.g., insufficient output at the actual system head).
2. Changes in suction or system pressure head that force the pump to operate inefficiently.
3. Clogging of pump suction, impellers, volutes, chambers.
4. Abrasion from silt or sand.
5. Corrosion due to improper materials for the environmental conditions.
6. Insufficient uphole velocity past submersible motors (common for small pumps in big casings with big drawdown), resulting in motor running hot.
7. Increased suction head from excessive drawdown, causing cavitation and pump-end damage.
8. Leakage into rubbed power cable or flexed connections, causing short circuits or stray currents.
9. Running dry (submersibles): overheating and distortion.
10. Clogged intake valves.
11. Particles lodged under intake valve or check valve seats, causing loss of prime.
12. Misaligned power-transmission rods or shafts, causing wear to bearings.
13. Overheating or freezing of exposed parts and motors.
14. Line power surges or brown-outs (capacitor, switch, and windings damage).
15. Improper line power or generator voltage, amperage.
16. Worn cylinder gaskets, seats, plunger parts (piston pumps).
17. Breaks in rubber components of bladder pumps.
18. Clogged intake or discharge lines (overheating likely).
19. Disruption of power supply from whatever source (nothing pumped).

Note: But it is amazing how well they work anyway. A more detailed set of pump troubleshooting charts are included in the Appendix.

be made to keep the air in the well column above freezing. Wellhead fittings should be covered in any case and can also be heated if necessary. Heated casing columns or wellheads should be vented to exhaust volatiles and humidity. Gas used for pump power should be as dry as possible to limit condensation in the air lines.

Portable pumps and bailers: For the purposes of well maintenance planning, portable pumps transported from well to well pose different types of long-term maintenance questions. The pumps are more accessible and not exposed for long periods to corrosive water. The main maintenance concern here is service under difficult site conditions, including abusive or careless use, and ease of repair and decontamination.

Decontamination is best conducted under ASTM D 5088. Total decontamination can be a problem under field conditions. For example, the instructions for the Grundfos Redi-Flo 2 pump recommend replacement of the water in the motor chamber between wells, a nine-step procedure (Grundfos Pumps Corp., 1992b).

Disassembly of most other pumps is hardly more convenient (not to pick on the Redi-Flo, but disassembly currently requires a special tool).

Some pumps can be completely flushed without disassembly. For example, progressive cavity designs with power from the surface (e.g., Keck Equipment Co., Williamston, MI) are readily decontaminated by reversing flow to empty and then pumping decon water through the system.

The quality of the pumped water has to be a consideration because water containing abrasive solids can clog or wear pumps and can be difficult to clean out. Oily water may cling to pump and line surfaces and require frequent detergent washing.

Mechanical matters also come into play. For pumps that are repeatedly inserted into and withdrawn from wells, it is important to pay attention to line abrasion and bending at cable or air line motor junctions. Field washing and decontamination can become a mechanical concern if deionized water freezes in the pump mechanism or lines (a problem below 20°F air temperature). In pump selection, if a particular pump type contemplated for project use is likely to become a maintenance problem, avoiding that type is step one in pump maintenance decision making.

Care must also be taken with electrically powered pumps that the power supply matches the characteristics of the motor. Check connections and generator or inverter output to make sure the right volts and amps are consistently being supplied. Manufacturers supply detailed service and troubleshooting instructions, which should be read and not just filed.

If maintenance and reliable operation are major concerns in the project being planned, bailers dedicated to individual wells have many advantages. Bailers have no moving parts aside from balls and stopcock valves. They require no power transmission via air line or cable and are not prone to clogging in routine use. Stopcocks and check valves may block open or closed with silt or sand.

Difficult or slow purging, sample aeration, contamination, and other factors may weigh against bailers from a sampling viewpoint, but from a maintenance angle, they are simple and reliable in operation.

Pumps in Remediation Extraction Wells

In the past, remediation wells were simply equipped with off-the-shelf submersible well pumps designed for water wells, and this is still often the case. Such pumps are designed to pump essentially clean water at reasonable efficiency with 10- to 20-year life cycles. Alternatively, dewatering-type eductor pump systems may be employed in very shallow conditions (Powers, 1992), especially for plume control.

Submersible well pumps have improved tremendously in service reliability in recent years. Corrodible materials have generally been replaced by noncorrodible plastics and stainless steel. Motors have improved service life and most are directly water cooled. The general-service submersible well pump is perhaps the most reliable device in the groundwater industry.

Eductor pump systems have no moving parts at the pump end, relying on jet action to aspirate water upward. Water power is applied by surface-mounted centrifugal pumps. Such systems have a lengthy history of service in dewatering projects, and geotechnical engineers designing similar contamination control projects naturally employed this type of system.

Still, not all such pumps do well in remediation wells. Silt and excessive metallic oxides are abrasive and quickly wear plastic impellers, seals, and bearings in centrifugal pumps. Corrosive conditions may exceed the designed limits of stainless steel components intended for use in circum-neutral pH well water.

The small volute channels of submersible pumps (even those designed and priced for environmental applications) provide a flow restriction where fouling deposits tend to accumulate (sometimes in days to weeks). Until there is some design breakthrough, this will continue to be a maintenance consideration.

Eductor pumps are vulnerable to clogging by metallic oxides. When water containing high levels of iron is exposed to the oxidation occurring as eductors pump air, clogging can occur rapidly. Iron biofouling can seal off eductors as well as the water circulation system very rapidly. This is a problem in shallow, organic-rich groundwater even without anthropomorphic organic contamination. Table 5 summarizes maintenance characteristics of common pump types.

Power systems are important considerations for extraction well pumps as well as those for monitoring wells. Power supply and control considerations relevant to environmental projects are discussed in detail by Powers (1992). Power to dedicated pumps should be consistent and secure. The voltage, amperage, and phase should closely match the requirements of the pump. Connections should be secure and weatherproof. Control boxes should be sealed from moisture and shielded from excessive dust, sunlight, heat, and cold (not always easy on a remediation site). Cables running to wells should be routed away from heavy traffic or excavation and protected from crushing. Lines should be clearly marked and respected.

On long-term projects, power systems and boxes should be vandal resistant. In potentially explosive atmosphere situations, controls should also be spark resistant and meet relevant code standards for this purpose.

Pump Protection

Once pumps have been selected based on their characteristics and installed, steps need to be taken to protect the pumps in operation. The two biggest problems are power supply and mechanical clogging and wear.

Power protection involves isolation from power surges (line or lightning), including abnormal fluctuations in both voltage and amperage. Motors should be securely grounded to an adequate dead ground separate from the well in order to redirect line surges and stray currents. Motors should be protected from running

Table 5   Maintenance Characteristics of Pump Types

Submersible centrifugals and turbines

Generally trouble-free and will work even when partially clogged; motors durable and reliable. Intake clogging, abrasion from sand/silt, "pancake" impellers easily clogged by Fe and S biofouling by-product encrustation. Corrosion of stainless steel under high-TOC, low-redox, or high chloride conditions, abrasion and flexing of power cable/connections, motor locking and overheating under some conditions, windings/capacitor burn-out from line surge.

Assessment: Pumps constructed of stainless steel with Teflon seals and bearings, with the most open possible impeller and volute channels are optimal for recovery and plume control. Of the following, the only really practical design for this purpose below suction-lift depths. Install with cable protection on easily pulled discharge pipe for cleaning.

Gas-driven bladder pumps

Generally low maintenance, portable, easy to clean. Will pump silt and sand, and will pump dry without damage, bladders and valves eventually susceptible to abrasive wear, intake clogging due to encrustation, supply-air line freezing. Permanent full-casing installations (grouted in) are difficult to retrieve and repair unless designed with removable elements.

Assessment: Well-suited, low-maintenance purge and sampling pump for monitoring wells, when properly selected and installed, maintained, and supplied with dry air. Permanent installations should be designed for removal of pump working parts.

Helical-rotor electric submersible

Generally easy to use, portable, low maintenance, even with silty, turbid water. Difficult to disassemble and clean, but backwashable. Motor leads susceptible to flexing and leakage when abused. Overheating a problem under prolonged pumping.

Assessment: Good, "high"-volume portable sampling pump when regularly maintained.

Force-actuated (Waterra) pump

Extremely simple, with no downhole rotating parts, all motorized components above-ground. May be hand- or motor-powered. Particles may lodge in check valves.

Assessment: Low maintenance, field durable, easy to clean mid-volume purging and sampling pump.

Push-rod piston pumps

Positive displacement, will pump dry without damage, simple mechanical components. Motor parts above-ground. Check valves, seats, plungers susceptible to abrasive wear and may be blocked by larger particles.

Assessment: Suitable for high-head, low-volume purging and sampling analogous to oil well stripping. High maintenance when pumping abrasives, otherwise low. Occasional plunger rebuilding. Rods must be noncorrodible (fiberglass available).

Suction-lift eductor pumps

Dewatering-style eductors provide positive lift to suction-lift centrifugals. Easily serviced, low-cost, high-volume pumps. Will pump air without damage. Susceptible to eductor and return-water line clogging by biofouling, encrustation, and silt.

Assessment: Highly suitable for shallow, high-volume pumping such as for dewatering, plume recovery. Eductors and lines on long-term systems must be cleaned frequently if pumping oxidized water. Relatively trouble-free when pumping anoxic water.

Note: This is a selection of more commonly used remediation, plume-control, volume-purging, and sampling pumps.

dry or hot. If loss of submergence or flow along the motor is likely, they should be equipped with sensors to cut off power if temperature or flow vary from established norms.

Especially where screens cannot fully prevent immigration of particles, pumps need to be protected against sand and silt pumping to limit wear. There are a number of approaches to take.

Including an integral suction flow control device (SFCD) (e.g., Pelzer and Smith, 1990; Nuzman and Jackson, 1990; Ehrhardt and Pelzer, 1992) in the well intake design of pumping extraction wells prevents sand and silt pumping and also buys time before screen performance decline due to encrustation begins.

The SFCD is an engineered pump tailpipe (for both submersible and lineshaft turbine pumps). It extends down to the bottom of the screen, and all pumped flow is forced through it. The SFCD intake perforation profile makes it more hydraulically open at the bottom of the intake pipe than at the top. The upper part of the screen is where most flow tends to enter the screen in a conventional pumped borehole well because the pressure is lower near the pump intake. The effect is to force a more cylindrical flow into the well (Figure 15).

SFCD have a demonstrated track record in minimizing sand and turbidity, even under difficult conditions. SFCD can also serve to minimize encrustation effects by making the entrance velocity more uniform.

In some very shallow wells and in other situations where an SFCD may not be suitable (where the ratio screen length to screen diameter is <30), centrifugal sand separators can be mounted on the well pump to remove sand prior to entering the pump intake (Figure 16). These discharge sand to the bottom of the well, but some field experience suggests that it ceases to pile up after a hydraulic equilibrium is reached (Lehr, 1985). One drawback to discharging sand to the bottom in pumped wells that are monitored for quality is that particulates may hold and intermittently desorb constituents.

Design and selection is ideally where a monitoring and remediation well project should start. Maintenance takes over as the operational issue once the system is constructed. This is considered next.

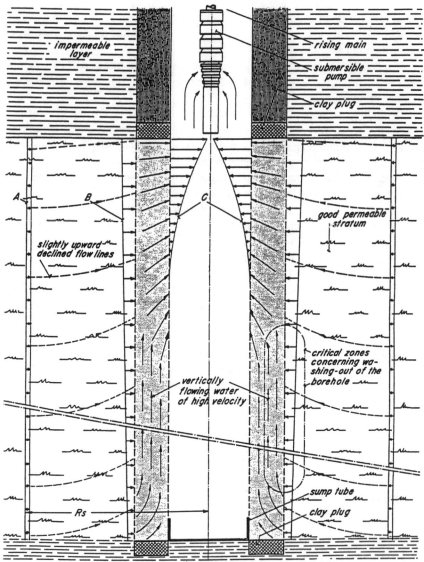

A · velocity profile of afflux at defined radial distance Rs
B · velocity profile (acute triangular) of inflowing water while leaving the stratum
C · velocity profile of inflowing water within the slots of the screen

(a)

Figure 15.   Wells with and without SFCD: (a) Flow pattern in a convention-
ally constructed gravel-packed well; (b) flow pattern in a well
equipped with a SFCD. (Pelzer and Smith, 1990.)

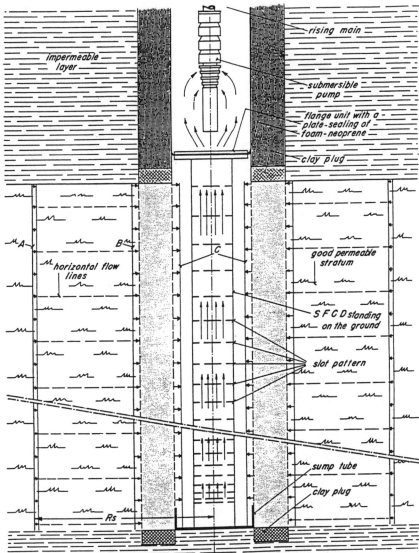

A: velocity profile of afflux at defined radial distance Rs
B: velocity profile (rectangular) of inflowing water while leaving the stratum
C: velocity profile (rectangular) of inflowing water within the slots of the screen

(b)

Figure 15.   (Continued.)

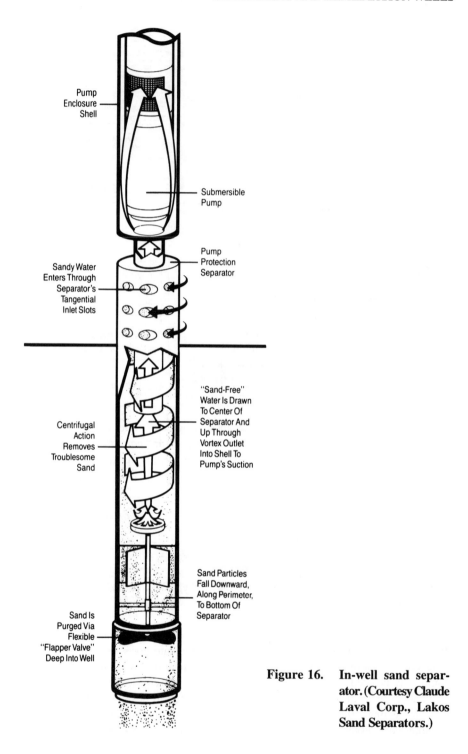

Pump
Enclosure
Shell

Submersible
Pump

Pump
Protection
Separator

Sandy Water
Enters Through
Separator's
Tangential
Inlet Slots

"Sand-Free"
Water Is Drawn
To Center Of
Separator And
Up Through
Vortex Outlet
Into Shell To
Pump's Suction

Centrifugal
Action
Removes
Troublesome
Sand

Sand Particles
Fall Downward,
Along Perimeter,
To Bottom Of
Separator

Sand Is
Purged Via
Flexible
"Flapper Valve"
Deep Into Well

Figure 16.   In-well sand separator. (Courtesy Claude Laval Corp., Lakos Sand Separators.)

5   Environmental Site Well Maintenance: Issues and Procedures

Once wells are designed, installed, and in operation, they must be maintained to prevent or slow deteriorating conditions. Again, if you need a cure, go ahead and make it happen, but come back here to see how to avoid a recurrence.

Once design, construction, and development are completed, well maintenance has to begin, based on a preconceived maintenance plan regularly modified to fit the field conditions. Despite design and care in construction and development, aquifer water quality problems in environmental situations are inevitably difficult. Future problems with wells are likely to occur and just have to be planned for and dealt with.

There is a cost to preventive maintenance in operator time, contractor and consultant assistance, maintenance contracts, spare parts and equipment, analyses, and record-keeping. The cost-benefit decisions on well maintenance depend on the local situation, but studies and experience have shown the following general relationship for municipal water wells: preventive maintenance costs are 10 to 20% of rehabilitation, which are 10 to 20% of new well construction (Smith, 1990) (Figure 17). Figure 18 illustrates a decision-making flowchart illustrating some of the factors involved in making rehabilitation vs. maintenance cost decisions.

In monitoring and recovery well planning, the direct preventive maintenance costs, rehabilitation costs, and new construction are at first glance relatively more similar in cost. This is due to the relatively higher costs of maintenance and rehabilitation and the relative inexpensiveness of well construction in shallow aquifers by hollow-stem auger.

It is important to remember that considering direct costs of various options alone can provide a distorted picture. Direct, immediate costs tend to exclude some real operational expenses and also the implications of future problems temporarily avoided. For example, rehabilitation and new construction involve the personnel expense of submitting rehabilitation plans or new designs to the authorities, the meetings, the questions, etc. Maintenance minimizes the need for such planning work, especially if the maintenance plan is part of the initial site plan.

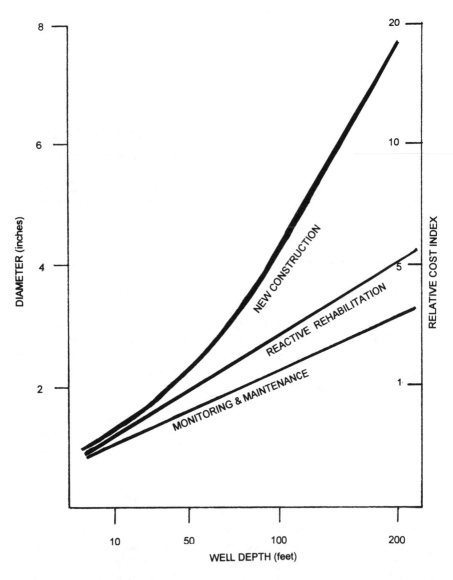

**Figure 17.   Theoretical relative cost effectiveness of well maintenance, re-
habilitation, and new construction.**

New well construction may simply serve to temporarily avoid recurrence of
a problem. Current experience is demonstrating that clogging, biofouling, and Fe/
Mn/S transformations may extend several meters away from existing problem
wells (Figure 19).

The need to control or monitor plumes usually constrains where new wells
can be placed. They consequently often end up within the "problem zone" of the
abandoned problem well and subsequently rapidly fall to the same symptoms.

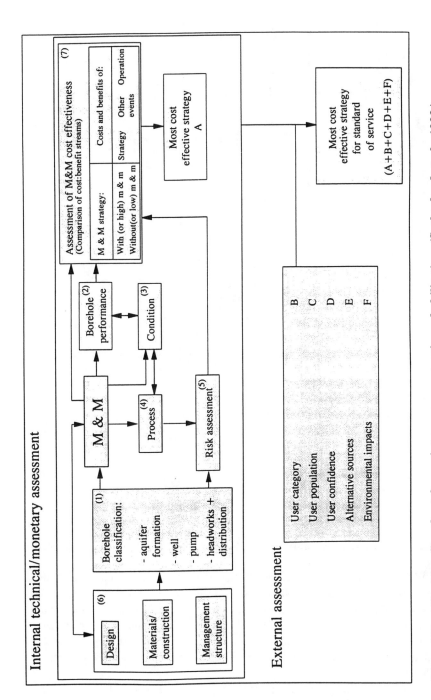

Figure 18. Decision flowchart for maintenance vs. reactive rehabilitation. (Sutherland et al., 1993.)

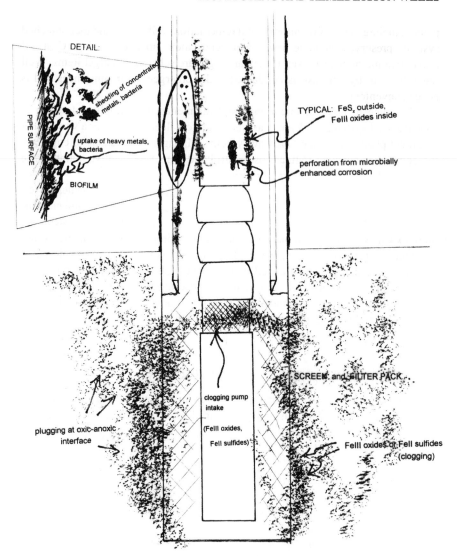

Figure 19. Fe transformation and plugging zone around an affected well.

Well maintenance is therefore a better option than abandonment and new construction when existing well systems are well designed and constructed because the old problem cannot be easily avoided. Where existing failing wells are not up to current standards and are detailed for abandonment, the performance problem of the former wells cannot be considered solved by installing new wells. Well deterioration will recur with the new wells unless a maintenance program is implemented.

Finally, regardless of comparative costs, the intrinsic value of smooth operation has to be considered when making management decisions about maintenance. As in the operation of any engineered system (e.g., trucks or nuclear power plants), maintenance of monitoring and remediation well arrays and their attached systems preserves valuable assets and keeps them up and running. Costs of preventive maintenance, even if they are high, are scheduled budget items. Well rehabilitation, by contrast, is conducted under emergency circumstances and is never convenient.

A choice in favor of intentional well maintenance is simply a decision to consider well systems as physical assets, just like a fleet of trucks or a wastewater treatment plant, rather than throwaway items. Management systems of engineering companies consider regular cost control and regular preventive maintenance as part of the cost of operating a physical asset.

The principles are precisely the same for wells as for other mechanical assets that are expected to work without much attention in a hostile environment: none works well if it is neglected. Like vehicles and wastewater plants, well systems are more expensive to fix and replace than to maintain properly over their life cycles.

Unlike vehicles and wastewater plants, problems with wells are not visibly apparent. Rational, systematic well maintenance is, however, entirely possible.

Routine well maintenance has a short engineering history; but where systematic maintenance inspection, monitoring, and treatment has been employed, they have visibly reduced clogging and other impacts on well performance, piping, and filtration and treatment equipment.

MAINTENANCE PROCEDURES OVERVIEW

Wells are best maintained by a preventive maintenance program that involves a combination of regular monitoring of their physical condition and well performance factors, and reconstructive maintenance and preventive treatments as necessary.

- *Maintenance monitoring* is monitoring of well physical, hydraulic, and water quality factors for the purpose of detecting deterioration conditions.
- *Reconstructive maintenance* is replacement of well and pump components on an as-needed or scheduled basis.
- *Preventive treatments* are redevelopment or chemical treatments applied before performance decline occurs, often on a schedule once the rate of corrosion, biofouling, etc., is known.
- *Rehabilitation* is the next resort when a well's performance is allowed to deteriorate too far. The last step is abandonment and starting over. If well failure and replacement has had to happen in the course of events, then the well system manager has a chance to "get religion," implementing a maintenance program to prevent or delay a recurrence.

Implementing a Maintenance Program

The primary task in a preventive maintenance program is, first of all, to make the institutional commitment to perform preventive maintenance on wells, assuming that well deterioration is a potentially serious and costly problem. It is hoped that the case for this assumption has been amply made here. It also may be notable that ASTM well maintenance standard guides are under development, meaning that a lot of people think that maintenance is a good idea for environmental wells.

The commitment has to be institutional and not the personal crusade of one facilities manager, who may die, retire, or otherwise move on next month. If it is the personal commitment of one person only, the momentum toward a maintenance program is lost in the transition until the next manager encounters a crisis.

Maintenance Basics

As in any kind of maintenance, well maintenance includes some simple asset-protecting activities. Among these are simply knowing where wells are, making sure they are accessible and visible (and not buried in a cell somewhere), and checking on their surface equipment. Are the caps on? Are the locks working? Are the well tops visible and properly labeled? Are their well construction (as built) records available? Do pumps operate within nominal ranges?

Well System Maintenance Records

Well records are among the most valuable tools in well maintenance because they provide the necessary dimensions and history to make maintenance and rehabilitation effective. Without records, M&R are much less certain of success and more prone to failure.

In addition to construction data, the maintenance system should have an accurate record of well operational information and data, such as water quality, recorded over time. These records should be organized in a logical fashion within the framework of the site's management system and readily accessible to people in the organization concerned with well maintenance.

Such a record system may consist solely of hard-copy paper files or may include a database-spreadsheet system, such as is presumably available to the vast majority of managers of monitoring and remediation programs. The computerized system speeds data retrieval and analysis. Specific well system management software is available (but is likely not to exactly fit every facility's needs without modification). Business-type spreadsheet-database software may suffice and can be adapted by knowledgeable computer people to fit the well maintenance needs of the facility.

Computer systems should be augmented by hard-copy or videotext files for video tapes and documents such as well logs, invoices for services, and test results.

It is important that maintenance records be available to all the relevant parties on a need-to-know basis. Site owners should have contractual access to records

Table 6 Well and Pump Monitoring in Well Maintenance

- *Maintenance monitoring*
  Routine performance checks conducted as a part of the scheduled mainte-
  nance program of the water pumping and treatment and/or data acquisition
  system. Goal: to spot changes and deterioration in performance.
- *Diagnostic testing*
  Diagnostics are performed to evaluate the causes of deterioration in wells
  (once it has occurred) in order to select an action to improve the well if
  possible.
- *Post-rehabilitation testing*
  Tests conducted to evaluate the effectiveness of rehabilitation.

maintained by contractor management and well maintenance firms. These files
remain the property of the site's responsible party in case of changes in contractor
firms (hardly an unusual occurrence).

Maintenance Monitoring for Performance and Water Quality

Maintenance monitoring is the process of performing systematic monitoring
to permit early detection of deterioration that may affect the well's hydraulic
performance and water quality. The ideal is to detect deteriorating effects in time
to prevent problems or allow the easiest possible treatment. Table 6 summarizes
purposes of monitoring and Table 7 summarizes factors of interest in maintenance
monitoring.

The same evaluations can be applied in principle to monitoring wells. Perhaps
hydraulic performance changes are not so noticeable, but changes in or variability
in sample quality (constituents and turbidity), and static and end bailing/pumping
levels, can be determined during testing.

Maintenance Actions and Treatments

Once maintenance or diagnostic testing has indicated that a deteriorating
condition is likely to cause a problem, or has established a suitable maintenance
interval, some action needs to be taken. This may be in the form of an inspection
and repair of a component such as a pump, replacement or cleaning of a filter, or
some treatment such as redevelopment, designed to minimize or correct a problem
condition. Some actions are considered in subsequent chapters.

Table 7   Performance Checklist

This checklist can be used to evaluate the performance of a pumping well. Ask
What is the:

- Normal pumping rate and how many h/day (or sampling event, mo., yr.) does it operate?
- General trend in water levels in wells in the area or on the site?
- Amount of drawdown in the well that results from drawdown interference of nearby pumping wells?
- Normal pump motor voltage and amperage and what is it now?
- Total depth of the well and its physical structure (and has it changed)?
- Static water level in the production well (now and in the past)?
- Pumping rate after a specified period of continuous pumping (now and in the past)?
- Pumping water level after a specified period of continuous pumping (now and in the past)?
- Specific capacity after a specified period of continuous pumping (now and in the past)?
- Well efficiency in volume pumped per kWh input (now and in the past)?
- Sand or other particulate content (e.g., turbidity, NTU) in a water sample after a specified period of continuous pumping (now and in the past)?
- Water quality (physicochemical, biological) of the well and how has it changed over time?

Note: The key factors are not really the actual number values, but the changes over time. A significant change in any of these conditions indicates that the well is in need of attention.

6   Maintenance Monitoring Program for Wells

Monitoring is a key part of preventive and proactive maintenance. This section describes decision-making, considerations, and recommendations for practical preventive maintenance monitoring.

Managers of monitoring and remediation well systems (like those of any engineered system) make operational decisions based on formal or informal cost-benefit analysis. Deciding how to maintain a system properly requires recognizing the risks to the system. Recognition requires knowing what to look for, such as those factors outlined in Chapter 2.

Assuming that monitoring and remediation wells will experience a variety of problems, a variety of risks have to be evaluated. It has been established that deteriorating conditions in wells can be complex and best controlled if detected early. For these reasons, effective maintenance has to be based on regular monitoring, including electromechanical, physical, chemical, and microbial factors, as well as pump and well service and record-keeping. Because of its importance, monitoring is considered here in some detail.

PURPOSES OF MAINTENANCE MONITORING

Maintenance monitoring of such parameters provides useful knowledge of the nature of problems, which permits reasonable countermeasures to be explored. Repair and replacement intervals and preventive maintenance treatments can then be chosen and fine-tuned accordingly. Effective maintenance of wells without monitoring of well parameters is no more possible than effective vehicle maintenance without analysis of engine operation and regular mechanical inspection.

The effects of chemical action and solids such as silt on well performance, and methods to monitor for them, are typically rather well known to managers of environmental site well arrays. Monitoring for these factors should be relatively easy to sell if there is any commitment to maintenance whatsoever. Biofouling monitoring for wells has only recently gained rational sampling and suitable analytical methods. Unlike water level, flow, physicochemical and silt and turbidity analyses, biofouling diagnosis methods are still not standardized (and may never be). For these and other reasons, biofouling monitoring is not as well appreciated.

However, biofouling and changes in physical parameters, such as turbidity and sand content, among all the indicators of deteriorating effects, are the most amenable to preventive or early warning monitoring. To be treated effectively, all have to be detected as early as possible. Such monitoring permits making reasonable judgments, for example, of how quickly biofouling is occurring, its effects on the system, and how it can be controlled. (Why Fe, S, and Mn biofouling is a particularly vexing problem for well maintenance is discussed in Chapter 2.) The control of sanding and silting also benefits from early warning so that troubles can be tackled early on, before a pump is ruined or a screen collapses.

The recommendations and rationale presented here are designed to be used in presenting a convincing argument to site management that well monitoring, including biofouling factors, should be budgeted as a part of a maintenance program designed to protect capital assets and provide the best possible system performance.

The following recommendations are primarily based on the experience of consulting and research projects, including that which culminated in the AWWA Research Foundation publications of Smith (1992) and Borch, Smith, and Noble (1993), but also other published reports (e.g., Howsam and Tyrrel, 1989; Cullimore, 1993) and unpublished consulting projects to date. Attempting to maintain systems without such monitoring is virtually pure guesswork — gambling on short-term savings without much hope for any payoff in any currency but grief later.

No monitoring program can prevent deteriorating conditions from occurring. However, with such a monitoring program, taking effective countermeasures is possible, resulting in long-term savings.

At a minimum, a preventive monitoring program should provide for regular analyses to determine: (1) whether a deteriorating condition may be occurring, and (2) the reasons for changes in well and pump performance and water quality as soon as they can be detected. To make use of such information over time, a maintenance system must have organized and accessible records.

DECIDING HOW TO MONITOR

The recommendations of this section should be considered guidance in making decisions about a biofouling monitoring program rather than a standard guidance of methods or detailed manual of action. Both the monitoring tools available and the methods of employing them are evolving too rapidly to be formalized into a definite standard procedure. Besides the recommendations here, there are alternative methodologies, essentially variations on a theme; for example, biofouling monitoring methods presented by Cullimore (1993).

Selection of a monitoring program should proceed based on a thorough technical assessment of the wells of interest. At a minimum, system operators should monitor: (1) hydraulic performance, including pump and motor characteristics (voltage, amperage, vibration); (2) physicochemical parameters relevant to the system; and (3) indicators of growth and/or occurrence of biofouling micro-organisms in wells. Methods chosen should be as consistent as possible over time, but allow for changes as appropriate.

The maintenance monitoring recommendations presented are designed to detect a variety of conditions and symptoms, many of which are interactive. For example, consider a circular interaction situation in which silting is aggravating biofouling clogging, which is reducing the hydraulic efficiency of a well, resulting in initially increased drawdown, and then apparently recovers when the pump subsequently wears and clogs, thereby reducing yield. High motor amperage draw with low flow output signifies pump clogging, while low amperage draw and low output indicates a leak somewhere (perhaps indicating corrosion).

The recommendation contains a heavy emphasis on biofouling analysis because it is the most troublesome and common problem in wells and the most difficult to rectify effectively after the fact. Monitoring of hydraulic performance and solids such as sand are equally important in most systems. None of these factors in well deterioration can be profitably neglected.

The following describes some specific monitoring recommendations for the detection and monitoring of deteriorating conditions in wells. Recommendations are made for physicochemical, biological, and hydraulic-performance monitoring methods. These are further classified into two levels of complexity for operational use. Selection of a level, or combination of levels, should be based on an analysis of the information the methods can provide, as well as the technical needs and financial resources of the facility.

A separate budget analysis serves to determine the long-term cost benefits of maintenance monitoring and preventive treatment vs. reactive rehabilitation. The cost analysis methods presented by Helweg, Scott, and Scalmanini (1983) and elaborated by Borch, Smith, and Noble (1993), although simple, provide a basis to do this for pumping wells. A new system (also developed for water supply well systems) is also presented. This spreadsheet-based system can provide rational comparisons among a variety of alternatives such as monitoring and maintenance (M&M) vs. doing nothing and levels of M&M. The cost analyses specific for monitoring well maintenance are yet to be definitely developed.

Since facility managers and operators are likely to be inexperienced with both the causes of well deterioration and methods for its monitoring and control, seeking outside expert help in getting started is highly recommended. Fortunately, it so happens that there is a community of professionals who are well experienced with these specific methods, their benefits, and limitations. Among these are authors of many of the references cited. Ideally, once properly implemented with adequate training, well maintenance monitoring programs should proceed without outside help unless the facility managers wish to subcontract the M&M.

MONITORING METHODS OVERVIEW

The following is a survey of existing monitoring methods and their uses in preventive maintenance and diagnosis of problems.

Visual Examination

Visual inspection of the well and pump components is a valuable adjunct to pump testing and analysis and can reveal important signs of corrosion and encrustation. For better or worse, pumps and pipe components in systems experiencing well deterioration serve as high-cost coupons or sacrificial indicators of corrosion and encrustation, such as those illustrated in Figure 20.

Another form of visual inspection is by borehole video camera. Borehole video recording provides direct visual information on the well and can be used to record changes over time. High-resolution color reception is highly useful and becoming available with camera heads of less than 2-in. diameter. The most preferable equipment is that which provides a right-angle color view that permits direct observation of the casing and screen. Figure 21 illustrates a typical borehole television system (Laval Underground Surveys, Anaheim, CA). More important than sophistication in equipment is a knowledgeable camera operator who can interpret accurately what the camera reveals.

Although the purchase of a borehole video camera or the hiring of its services represents a significant cost, the amount of information that can be obtained in a brief survey makes the survey a good value. Presumptive identification of borehole and casing wall fouling, encrustation, corrosion, and other structural damage can be made rapidly through downhole inspection by the experienced observer. These can be confirmed by direct testing of deposits or equipment retrieved.

Well and Pump Performance

Well hydraulic performance monitoring should consist of regular (e.g., monthly or quarterly) measurements of the following pump operation and hydraulic characteristics:

1. Pumping dynamic water level (DWL) (pumping water level, PWL) in pumping wells.
2. Static DWL ("true" static water level [SWL] or static DWL at the well with the pump off but under the influence of nearby pumping wells) for all wells.
3. Wellhead pumping rate and operating hours for regularly pumping wells.
4. For regularly pumping wells, pump power consumption and power characteristics (voltage, amperage draw, occurrence of stray currents [ohms]), especially noting how each varies from the manufacturer's nominal specifications.

The pumping rate should be determined against a consistent system head (if pumping into a collection system) and periodically against free discharge if feasible.

Figure 20.   Examples of visible indications of well and pump deterioration.

Figure 21.   Borehole video system for well inspections. (Courtesy Laval Underground Surveys.)

Methods of hydraulic pump testing and motor characteristics tests are widely known and referenced, for example in Borch, Smith, and Noble (1993) and Water Systems Council (1992).

Slug tests can also be used to detect changes in hydraulic conductivity for low-production monitoring wells as long as there is a historical record. If changes in conductivity are noted and there are no other hydrologic reasons, fouling of the well may be occurring. For arrays of monitoring and pumping wells, such as those used for plume control, automatic sensor-recorder methods make excellent sense in providing a lot of information with minimal operator labor or contact with contaminated fluids (if present). Table 8 summarizes flow and drawdown measurement methods.

Site managers of environmental projects typically have access to groundwater modeling capability with graphical output. Drawdown data from water level monitor output files can be input to the program for the well array and anomalies such as increased drawdown at particular wells can be displayed.

Physicochemical Analyses

The purpose of physicochemical monitoring for maintenance is to detect changes in parameters that may indicate conditions that cause or reflect well deterioration. This monitoring is separate in purpose from regulatory monitoring and has a different agenda. The needs are for more immediate results, more data

Table 8   Methods and Uses for Flow and Drawdown Measurements

1. *Flow:*

Temporary — Calibrated bucket (<10 gpm or 0.6 l/s);
orifice weir (>10 gpm) to open discharge (step tests).
Permanent/Regular — Turbine or venturi flow meter at well discharge;
Portable or permanent Doppler acoustic flowmeter. Electronic recording
recommended.

    Assessment: (1) Temporary methods are used to test new pumps or
retest existing pumps vs. open discharge. Both bucket and weir require
calculation or tables; margins of error are real but insignificant. (2) Permanent flowmeters are relatively expensive but cost effective in monitoring
flow. Wellhead measurement eliminates pipe-loss effect and allows individual well monitoring of multiple wells. Meters should be inspected and
cleaned periodically or they lose accuracy. Portable Doppler meters (expensive) are cost effective for multiple wells with no prospects for installing in-
line meters. Skill levels moderate.

2. *Drawdown:*

Air line, electric sounder, or acoustic (temporary)
Recording transducer method (permanent)

    Assessment: Drawdown and flow measurements together are required
to monitor well performance. Air lines or (better) transducer level sensors
for otherwise inaccessible wellbores. Air lines susceptible to clogging and
inaccuracy; electric sounders ubiquitous on groundwater remediation
projects: portable, accurate, require clear path to groundwater level; acoustic
level detectors for inaccessible wellbores: expensive, relatively inaccurate.
Transducer recorders permit accurate, automatic measurements, expensive
but reliable, and well adapted to automation. Internal drawdown tubes or
piezometers are recommended for accurate, safe measurements with electric
sounder. Skill levels minimal to moderate.

in a given length of time, and less need for rigidly certifiable, legally defensible
analytical precision. The need is to detect change over time early enough to make
decisions about maintenance.

    Basic water chemistry analysis should include at a minimum: soluble (Fe^{2+})
and total Fe, total Mn, total S^{2-}, pH, Eh, and temperature. Sulfate and S^0 or SO_4
solids monitoring may also be important on certain systems. Redox potential is
very important, both to the make-up of the microflora in the well and aquifer and
also to the fate of Fe, S, and Mn at the well, such as the mineral forms of
precipitates. Redox may be measured directly using appropriate electrode methods or estimated based on the redox couple ratio $Fe^{2+}:Fe^{3+}$, which is the only
relevant reversible ratio in ambient groundwater. It is notoriously difficult to
decide the meaning of a particular redox reading taken out of context; but charted
over time, patterns can emerge (e.g., Smith, 1992).

    Parameters relevant to formation of encrustation (e.g., Ca^{2+} ion) should be
determined. *Standard Methods for the Examination of Water and Wastewater*

Total Fe Data from Field Analyses

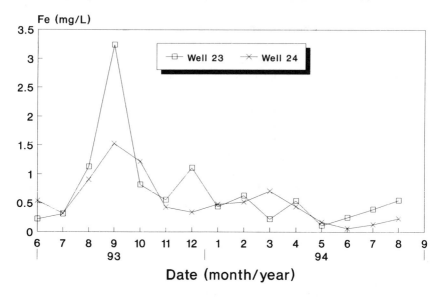

Figure 22.   Example physicochemical data chart.

(APHA, AWWA, WEF, 1992) provides guidance. Also, sediment content and type in bailed or pumped purge and sampling water (e.g., by ASTM standard method D 3977) are important gauges of screen and pack performance. These results can be charted over time, as for example Figure 22.

Total suspended solids can also be detected and measured using readily available turbidometers and particle counters. Turbidometers are the most widely used method for monitoring the particle removal efficiency of filters and they have wide commercial use. *Standard Methods* Section 2130 (APHA, AWWA, WEF, 1992) provides an instrument and method standard.

Particle counters are the subject of significant recent research. For example (again citing water supply research), Hargesheimer, Lewis, and Yentsch (1992) studied how particle counters can be used in monitoring water systems. Figure 23 illustrates particle counter configurations. Lewis et al. (1992) provide a useful description of particle counter use in practice. Such particle counters and turbidometers can monitor particle density automatically to save on sampling and analysis time, and have a second purpose in monitoring the performance of filtration.

Particle counters provide more information than turbidometers, in that they can both count and determine the size of particles. For surface water supply filtration, the size is important in detecting the possible presence of protozoan spores (Hargesheimer, Lewis, and Yentsch 1992; LeChevallier and Norton, 1992). In maintenance monitoring for wells, the emphasis again is on change in particle density over time. Important data are changes in particulate concentrations and characteristics such as mean diameter.

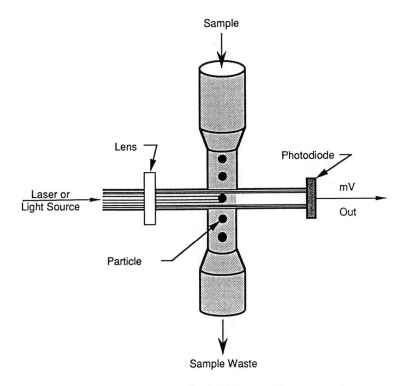

Figure 23. **Particle counter sensor configuration. (Reprinted from *Evaluation of Particle Counting as a Measure of Treatment Plant Performance*, Copyright 1992, American Water Works Association. With permission.)**

Cullimore (1993) describes a modification of particle analysis procedures that aids in the identification of particles; for example, whether they are mineral or precipitant in origin and what the precipitants might be. Such pigmented particles may be trapped from a known volume of water on a 45-μm membrane filter, air dried, and analyzed spectrophotometrically. Such solids data are useful in specifying remedial treatments.

As with microbiological methods (described in the following), particle counting and turbidometry are highly site specific in application (LeChevallier and Norton, 1992). Particle counting results, for example, cannot be compared among differing sites. However, a history of readings can be built to provide a history of a particular well site, which can be interpreted. As with microbiological data, to build such a history requires good records. Table 9 is a summary of physicochemical methods relevant in well maintenance.

Biological Monitoring: Decision-Making

A biofouling monitoring system should be chosen and implemented as part of the overall preventive maintenance monitoring program. Even when the problem

**Table 9   Summary of Physicochemical Methods
Relevant to Well Maintenance**

Fe (total, Fe^{2+}/Fe^{3+}, Fe minerals and complexes):
  Indications of clogging potential, presence of biofouling, Eh shifts.
Mn (total, Mn^{2+}/Mn^{4+}, Mn minerals and complexes):
  Indications of clogging potential, presence of biofouling, Eh shifts.
S (total, $S^{2-}/S^{0}/SO_4^{2-}$, S minerals and complexes):
  Indications of clogging potential, presence of biofouling, Eh shifts.
Eh (redox potential):
  Direct indication of probable metallic ion states, microbial activity. Usually
  bulk Eh, which is a composite of microenvironments.
pH:
  Indication of acidity/basicity and likelihood of corrosion and/or mineral
  encrustation. Combined with Eh to determine likely metallic mineral states
  present, and with conductivity and alkalinity to assess inorganic salts occurrence.
Conductivity:
  Indication of total solids content and a component of corrosivity assessment.
Turbidity:
  Indication of suspended particles content, suitable for assessment of relative
  changes indicating changes in particle pumping or biofouling.
Particle counts:
  Indication of suspended particles content, suitable for assessment of relative
  changes indicating changes in particle pumping or biofouling.
Sand/silt content (v/v, w/v):
  Indication of success of development/redevelopment, potential for abrasion
  and clogging.

is not obviously microbial in origin, microbial monitoring can often pinpoint root
causes. An example is $CaSO_4$ plugging, a secondary effect of sulfide oxidation
(which has a microbial component). As with other monitoring methods, biological
monitoring methods chosen should be within the technical and fiscal capacities of
the facility and expert technical support available.

The methods chosen should be capable of detecting indicators of biofouling likely
to cause difficulty for the well and water collection/treatment system at sufficiently
low levels to permit treatment before performance or water quality impacts become serious. Relevant microbiological methods are explained in somewhat more
detail here because of their relative lack of standardization and complexity in
selection and use, as compared to hydraulic and physicochemical methods.

Whether to Monitor for Biofouling

Whether to consider biofouling monitoring is a relatively easy question from
a practical standpoint. Fe biofouling in particular, S biofouling to a lesser degree,
and possibly Mn biofouling in some cases are apparently pervasive in many
aquifer settings. The literature indicates that there are no constraints on biofouling

in aquifers used for drinking water supply. Aquifers typically being monitored, or undergoing remediation, are also typically highly active biologically. These aquifers and the wells in them are almost certainly a setting for well biofouling.

In general, site managers and their consultants should consider "early warning" biofouling monitoring in any circumstances where they are: (1) monitoring shallow rock and unconsolidated aquifers, or (2) pumping contaminated groundwater or any groundwater with any measurable TOC for plume control or treatment. This may seem to be a sweeping recommendation, but there is a history of severe Fe, Mn, and S biofouling in such aquifer settings. A cursory literature database search, perusal of Smith (1992), Cullimore (1993), or Borch, Smith, and Noble (1993), or any of their references, and conversations with colleagues will provide examples of situations that may sound very familiar to the site manager or consultant faced with the clogging and loss of performance of their monitoring and extraction systems.

Biofouling Monitoring: What Methods to Choose

For biofouling, constant surveillance and willingness to act in response to indications of occurrence are essential. How the monitoring is carried out is important, but secondary to the primary requirements to commit to implement a useful monitoring program and to continue it over time.

A note about the current state-of-the-art in well maintenance monitoring methods: physicochemical water quality and hydraulic performance monitoring are relatively well developed — even standardized. They can be reasonably applied in a precise fashion and results, for the most part, are comparable from place to place.

Biofouling monitoring always (not sometimes) involves some degree of subjective judgment, both in implementation and interpretation of the results of monitoring. Current biofouling monitoring methods do not provide results that are quantitatively directly comparable to other methods of analysis, or other locations, or even other wells in a wellfield, although experience is providing statistical data that may bring a situation of comparable data (perhaps by the 20th ed. of *Standard Methods*).

This case-by-case situation must be clearly understood so that methods chosen are appropriate to the conditions in the wells in question and that each well is adequately monitored. Also, it must be understood so that those in charge of maintenance monitoring do not become frustrated with the methods.

The factors resulting in a particular analytical result are complex and interrelated and not yet fully understood. Significant change over time (e.g., months, seasons, years, or through a pumping cycle) provides some of the best insight on deteriorating processes taking place in a well or wellfield. To do this effectively, good records of data are required.

Biofouling Monitoring Methods: Analysis

Microscopic Examination and Analysis

Microscopic examination of water samples as well as fouling and encrustations can reveal stalk and sheath fragments of bacteria presumed to be involved in Fe, Mn, or S biofouling. Light microscopic examination has traditionally been the method of choice for confirming and identifying "iron bacteria." APHA, AWWA, and WEF (1992) and ASTM procedure D 932-85 document the commonly used procedures for sampling and analyzing samples by light microscopy for such so-called iron bacteria. (APHA, AWWA, and WEF [1992] Section 9240 [18th Ed.] on iron bacteria is in the process of being extensively modified for the 19th and 20th Eds.)

The optical resolutions necessary for observing bacterial structures are based on 400X to 1000X magnification (such as provided by a 40X common or 100X common or oil-immersion objective + 10X ocular) with a direct, filtered electric light source. Most microscopes are routinely equipped with maneuverable stage calipers, micrometer readings on the focusing knobs, and a micrometer scale in the ocular, and can be equipped with a camera attachment that permits the recording of observations.

Phase contrast and epifluorescent methods are useful enhancements for light microscopy, but not necessary for biofouling diagnosis, as most biofilm samples from wells provide good visual contrast. Electron microscopy provides ultrastructural details of specimens but is purely a research tool in the study of biofouling at the present time.

However, in many instances, biofouling as a cause of well problems may be difficult to diagnose via microscopy alone, even with very good tools and skills. Microscopy tells very little about the environment of biofouling deposition inasmuch as (1) the biofouling root cause of a symptom, such as the secondary precipitation of calcium sulfate mentioned earlier (e.g., Hodder and Peck, 1992), may be overlooked; and (2) samples examined may not include the filamentous or stalked bacteria normally searched for in such analyses. Samples may also not include enough recognizable materials to provide the basis for a diagnosis of biofouling.

Enhancements for the existing ASTM and *Standard Methods* procedures provide better results and aid in diagnosis. For example, Pedersen (1982) and Smith (1992), among others, describe collection of biofouling samples on immersed surfaces (glass cover-slips and slides) for analysis by light and electron microscopy. Other biofouling collection methods are also described in the literature, as summarized below. Alcalde and Gariboglio (1990) describe a simple enrichment and staining technique to enhance the numbers and visibility of filamentous bacteria under light microscopy. Cullimore (1993) recites some others. Bakke and Olsson (1986) describe an attempt to measure biofilm thickness by microscopy, but Tuhela, Smith, and Tuovinen (1993) found this type of measurement to be unreliable.

Other improvements in microscopic technique have been made to distinguish microbial components of biofilms from mineral components. It is also useful to note that examination of the particles of oxides themselves is a useful part of defining the nature of fouling in a well and water treatment system.

Culturing Methods

"Standard" Methods: Standard Methods (APHA, AWWA, and WEF, 1992) presents several formulations for nutrient media for heterotrophic Fe-precipitating

bacteria, Mn-oxidizing organisms, and *Gallionella* enrichment. In addition, numerous formulations are available in the microbiological literature.

Media for Fe-precipitating bacteria have been used with mixed success. No effort has been made to standardize these media with reference cultures from well water and thus the recovery efficiency of iron bacteria from groundwater samples remains unknown at the present time (Tuhela, Smith, and Tuovinen, 1993).

The *Gallionella* medium in *Standard Methods* is essentially based on Wolfe's (1958) formulation, which was subsequently modified by Christian (1975) and Hanert (1981). These media provide uneven recovery for unknown reasons, and full isolation of *Gallionella* is difficult but apparently possible with care (Hallbeck, 1993).

Because they do not seem to match up well with groundwater environmental conditions, the cultural enrichment media for heterotrophic Fe- and Mn-precipitating bacteria presented in APHA, AWWA, and WEF (1992) are seldom used among researchers working with Fe biofouling problems. Useful Mn-precipitation media are virtually nonexistent in common practice (e.g., Smith, 1992), although they are being refined (e.g., Emerson and Ghiorse, 1992). The available sulfur oxidizer media are still nonisolating, enrichment media.

Given their inconclusiveness, it is unlikely that a wellfield maintenance program would willingly devote the time, facilities, and effort required for these cultural enrichments, except in the interests of research. Fortunately, there has been some progress in making biological monitoring easier and more suitable for routine maintenance. Work is under way to improve and define these in future editions of *Standard Methods* (beyond 19th Ed.).

Prepared BART Methods: Currently, the most promising cultural approach for routine monitoring purposes is that developed by Droycon Bioconcepts Inc., Regina, Saskatchewan in Canada based on research first conducted at the University of Regina. These BART™ Method tubes contain dehydrated media formulations and a "floating intercedent device" (FID), which is a ball that floats on the hydrated medium of the sample. These devices and their proposed use are described in detail by Cullimore (1993). This method, which is gaining acceptance as a means of detecting biofouling microorganisms, was found by Smith (1992) in field trials to provide useful qualitative information in well biofouling events. Figure 24 illustrates the BART tube system.

The BART Method tubes come with a variety of media mixtures. The IRB-BART™, for example, is designed to recover microaerophilic heterotrophic Fe- and Mn-precipitating microorganisms. The IRB-BART has been presented as a method of detecting growth to provide a presence-absence (P-A) or semiquantitative (MPN) result (Mansuy, Nuzman, and Cullimore, 1990; Gehrels and Alford, 1990). The theory and recommended implementation of BART methods is elaborated by Cullimore (1993). In short, BARTs are interpreted based on: (1) visual appearance of the inoculated tubes at initial reaction and as reactions change, and (2) days of delay (d.d.) or time (usually days) until a noticeable reaction occurs.

The field tests documented by Smith (1992) support the previous work suggesting their practical use in maintenance monitoring and diagnosis, which is being further supported in practical use that is not well documented in public literature.

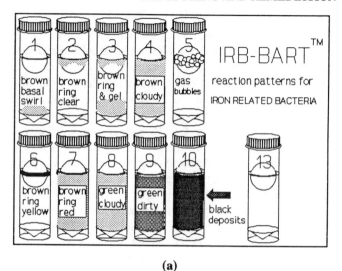

(a)

(b)

**Figure 24.   BART tube reaction patterns: (a) 11 reaction patterns for iron-
related bacteria presence; (b) SRB reaction patterns. (Cullimore,
1993.)**

The best use of BARTs is as a quick and easy indication of the: (1) existence
of probable heterotrophic Fe and Mn biofouling and the presence of sulfide-
forming SRBs, and (2) relative "aggressivity" or activity of the biofouling and
SRB communities.

They are, like *Standard Methods* formulations, enrichment media — not
precise research tools, but they can be unpacked and used under on-site conditions. This is their primary benefit as a monitoring tool.

With enough repetitions of samples, standardized sampling, and time, statistically significant data can be collected using these methods, which are useful in

**Table 10 Interpretation of SRB d.d. Data
as Aggressivity and Population**

Days of delay (d.d.)	Aggressivity	Possible population (log. cfu/ml)
1	Very high	4.1 ± 1.6
2	Very high	3.6 ± 1.4
3	High	3.4 ± 1.4
4	High	3.2 ± 1.2
5	High	2.6 ± 1.2
6	High	2.2 ± 1.4
9	Moderate	1.2 ± 1.2

D.R. Cullimore, unpublished data.

**Table 11 Interpretation of IRB d.d. Data
as Aggressivity and Population**

Days of delay (d.d.)	Aggressivity	Possible population (log. cfu/ml)
1	Very high	6.2 ± 1.4
2	High	5.4 ± 0.9
3	High	4.5 ± 1.2
4	Moderate	4.1 ± 1.2
5	Moderate	3.8 ± 1.4
6	Moderate	3.3 ± 1.4
7	Background	3.1 ± 1.5
10	Background	2.5 ± 1.2
15	Very low	<2.0

D.R. Cullimore, unpublished data.

making decisions about maintenance. For example, empirical field research by Cullimore and others have yielded the following tables (Cullimore, personal communication) (Tables 10 to 12):

Table 13 compares these BART data to plate counts made on R2A+FAC medium (R2A heterotrophic plate count medium amended) from Smith (1992). The values are comparable, but illustrate the range of results possible and the difficulty of comparisons with HPC data.

Biofouling Monitoring Methods: Sampling Methods

Grab samples can be collected by pumping, or biofilm material allowed to collect on in-well coupons or wellhead filtration devices.

Pumped Sampling

Pumped (grab) sampling is the easiest way to obtain samples from wells for analysis, including for evidence of biofouling. This method assumes that biofilm

Table 12 Community Structure Based on
IRB-BART Reaction Pattern Signatures

Reaction type no. pattern[a]	Community structure prediction
2–8–9	Mixed aerobic flora dominated by pseudomonads[b]
8–9	Flora dom. by pseudomonads
8–9–10	Ps. dominate with some enteric bacteria present
1	Covert[c] dense slime formers
1–4	Dense slime formers w/ mixed bacterial population
2–4	Mixed population of IRB
3	Enteric bacteria possibly dominated by *Enterobacter* spp.
2–3–4	Mixed IRB flora w/ *Enterobacter*
5	Deep-seated[d] anaerobic bacterial flora
5–4	Deep-seated anaerobic flora w/ aerobic IRB present
5–8–9	Deep-seated anaerobic flora w/ pseudomonads present
5–9–10	Deep-seated anaerobic flora w/ pseudomonads and enteric bacteria present
6	*Citrobacter* spp. may dominate
7	*Enterobacter* and/or *Klebsiella* spp. dominate flora[e]
2–5	Mixed aerobes w/ some anaerobic activity
2–5–6	Mixed aerobes w/ some enteric bacteria present

Note: Unpublished information provided by D.R. Cullimore. Readers are recommended to read Cullimore (1993) for further interpretive information.

[a] Reaction types are determined from comparison to a chart (see text). The numbers are the sequence observed over time as appearance of the tube changes.

[b] Pseudomonads are a highly diverse group of bacteria generally loosely assigned to the genus *Pseudomonas*, which are typically capable of great metabolic adaptation. Some are IRB and some are known opportunistic pathogens, also hydrocarbon degraders.

[c] "Covert" in Cullimore's usage is typified by occurrence away from direct observation; in this case, slimes formed in the formation away from the well.

[d] Deep-seated: occurring well away from the well.

[e] A number of known enteric types are known Fe precipitators.

bacteria and their characteristic structures are also present in the water column (planktonic phase). However, if pumping fails to detach and suspend biofilm particles, they will not be available for collection. The absence of bacteria in samples taken this way may simply mean that the bacteria remained attached, not that they are actually absent in the well.

Pumped samples may be analyzed by microscopy or by bacterial enumeration with selective or general-purpose nutrient media. Pumped water streams may also be evaluated using turbidometers or laser particle counters, both of which have use in automated monitoring, as previously described.

One limitation of pumped sampling is the "snapshot" nature of the samples, representing the water quality only at the time that the sample was taken. Shedding

Table 13   Comparison of Field HPC
(R2A+FAC medium) to BART Index

Days of delay	Range (CFU/ml, R2A+FAC)	BART index as log CFU/ml[a]
2	20–3000[b]	5.4 ± 0.9
3	<10–>2000	4.5 ± 1.2
4	10–>300	4.1 ± 1.2
5	10–200	3.8 ± 1.4
6	40–50	3.3 ± 1.4
7	20–80	3.1 ± 1.5
8	50–300	
9	300–1000	
10	30–1000	2.5 ± 1.2
12	20–200	
14	700–800	
15	≤10	<2.0
21	≤10	
22	80–400	

Note: Based on monthly and weekly samples from biofouled Ohio water wells.

[a] From Table 11.

[b] Data: combined HPC on R2A+FAC media. (Smith, 1992.)

Modified from Smith, 1992.

events may provide slugs that transiently increase microbial counts or the concentration of Fe and Mn in groundwater. Most bacteria in any single pumped water sample collected under these circumstances have been sloughed off the biofilm and are likely to represent only a tiny fraction of the population and diversity of organisms that comprise the biofilm.

After a period of sustained pumping, biofilms will yield very little of the turbid material usually necessary for microscopic examination. Analyses of samples taken after prolonged pumping may fail to detect the presence of chemical and microbiological parameters that would indicate the presence of biofilms near wells.

Cullimore (1993) describes a time-series, pumped-sampling program that attempts to overcome the "crap-shoot" nature of such grab sampling. Cullimore's procedures are divided into three increasing levels of complexity, from minimal and rapid to comprehensive and lengthy. All involve taking advantage of the phenomenon that biofilm sloughing occurs preferentially on start-up after a period of rest or "quiescence," in which the pump is allowed to shut down for a period of time from 2 h to several days. Samples are taken (1) just prior to shut-down, (2) immediately after restart, (3) a few hours later, and optionally (4) days later. Figure 25 illustrates some possible sampling patterns.

This approach, including taking replicates of samples at each sample event, helps to overcome the limitations of pumped grab sampling as the sampling

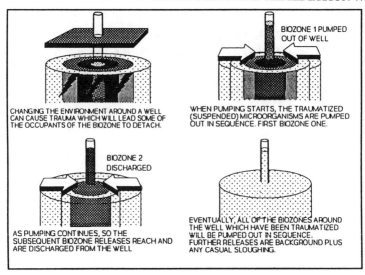

(a)

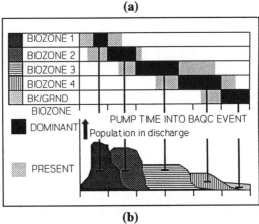

(b)

Figure 25.   BAQC tables and interpretation: (a) interpretation of presence of biozones based on BAQC interpretation; (b) schematic sequence of biozone discharges over time; (c) example BAQC days of delay (d.d.) data. (Cullimore, 1993.)

method for cultural analysis. Grab samples remain unreliable for microscopic analysis. For this, some method is needed to provide enough sample to view or otherwise analyze minerologically or chemically.

Surface Collection on Slides or Coupons

Surface collection methods provide a different procedure for detecting bacteria involved in biofouling phenomena in the wellbore and at the wellhead. These methods overcome some of the shortcomings of pumped grab methods in providing samples for microscopy or other analysis during regular well operation. While they also rely on the detachment and suspension of biofilm particles by pumping,

Before Quiescence (BQ); 2 mins into pumping (D1); 2 hours into pumping (D2)

BQ	D1	D2
*3	2	3
*4	3	4
*5	3	5
6	4	6
7	5	6
8	6	7
9	6	8
10	7	8
11	8	8
12	9	9
13	9	10
14	10	11
15	10	12

*wells would automatically be considered biofouled, values given are dd values for RX1 using the BART™ biodetectors.

(c)

Figure 25. (Continued.)

collectors have a longer time to gather particles. The result is that samples from collection surfaces provide essentially intact biofilms for analysis. This method is also adaptable for collection of samples of inorganic encrustations.

Collection surfaces can be placed at various locations in the water system. For collection of well biofilm samples, collectors may be placed directly in the wellbore or in the water pumped from the well. Several experimental designs for collecting biofilm organisms have been presented in the literature and are summarized by Smith and Tuovinen (1990) and Smith (1992). Cullimore (1993) additionally describes in-well incubation devices under development. The primary considerations for choosing in-well and/or wellhead collectors are (1) need for representative collection of biofilm samples, (2) access to the well for collection, and (3) influence of the collector on well performance.

In-well biofilm collectors are practical only if there is access to the wellbore itself through the well cap or nearby water-level monitoring wells (piezometers), and it is acceptable to introduce such a device into a well. An in-well collector has to avoid hanging up on submersible pump wire or other obstacles. The inserter devices refined by Smith and Tuovinen (1990) and Smith (1992) are designed for tight wells with submersible pumps and have good potential application in environmental wells.

Wellhead devices allow collection even on wells that are permanently sealed at the surface. Hässelbarth and Lüdemann (1972), Pedersen (1982), Howsam and Tyrrel (1989), Smith and Tuovinen (1990), and Smith (1992) have all described the application of sampling devices specifically to collect bacteria and well fouling material at the well head. Figure 26 is a schematic of the Smith (1992) wellhead flowcell collection device. Sample collection using this method will be described in the *Standard Methods*, 19th Ed., Section 9240.

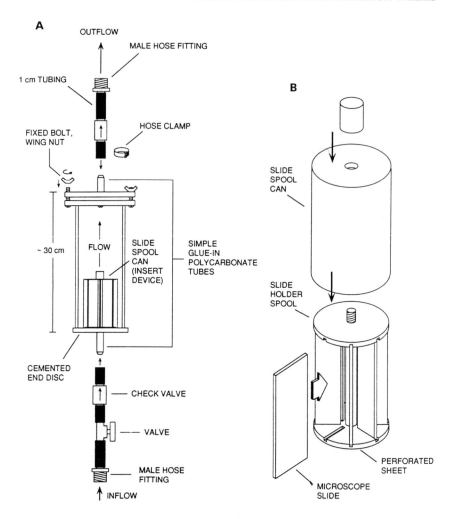

Figure 26. **Wellhead flowcell collection device: (A) assembly; (B) slide holder insert. (Reprinted from Smith, S. A.,** *Methods for Monitoring Iron and Manganese Biofouling in Water Supply Wells,* **Copyright 1992, American Water Works Association. With permission.)**

The flowcell system used by Smith (1992) with considerable success (and since refined) was a development of the 1990 design. It has proven to have the additional advantages of being readily field serviceable and adaptable for a variety of collection applications, as well as for use as a laboratory model (Tuhela, Smith, and Tuovinen, 1993).

The Howsam-Tyrrel moncell sand-filter collector (Howsam and Tyrrel, 1989) is an alternative collection device that has several advantages and disadvantages vis-à-vis the flowcell-slide system. Like the flowcell, a moncell can be observed directly and collects filamentous and other iron material. A moncell has to be subsampled and material extracted for examination, then renewed with new filter sand. This is a possible disadvantage, except that the moncell can act as a useful model for what can happen in downstream filters. Selection of one system or the other should be based on a comparison of cost effectiveness, needs and capabilities of the user, and availability.

Representativeness of Collection Sampling

Since biofilm communities are unevenly attached on immersed surfaces, there is a question of the representativeness of biofilm sampling. Glass and polycarbonate slides have been used for in-well collectors in a variety of work, all the way back to the late 1920s work of Cholodny (Hallbeck and Pedersen, 1987; Smith and Tuovinen, 1990). In addition, other surfaces may be used as well, including metals. This is an old technology in the collection of well samples for microscopy and is also used to collect surface water microflora, especially in streams.

One factor in collection-surface selection is the surface free energy of a surface, which influences bacterial attachment. Although their surface characteristics are somewhat different from those of stainless steel and other metal surfaces immersed in well water, glass slides are considered to be acceptable model surfaces because of the similarities of their surface free energy characteristics (Hallbeck and Pedersen, 1987).

In addition, slides are readily available, inexpensive, noncorrodible, and can be directly examined microscopically, or sampled for chemical analysis and for cultural recovery of microorganisms. However, collection of biofilm on materials used in the wells, analyzed by various means (including electron microscopy), would provide information on actual biofouling effects on materials used in wells.

The Howsam and Tyrrel (1989) moncell device, as an alternative collector method, collects deposits inside a sand-filled chamber from water trickling through the sand. In-line paper cartridge filters, such as often installed in pump-and-treat systems to protect carbon filters, can also be examined microscopically and subsampled for analysis of the biofilm present.

The collectors chosen in any case will collect a biofilm sample typical of that existing in the near-wellbore environment only. Wellhead collection (slide or filter) and pumped grab samples will also have been modified by the pressure and rotation changes exerted by the pump, resulting in samples that are likely to be somewhat more oxidized and physically disturbed than biofilms within the aquifer.

RECOMMENDATIONS FOR MAINTENANCE MONITORING IN ROUTINE PRACTICE

Based on the biofouling monitoring research of Smith (1992) and others (e.g., Cullimore, 1993), the following procedures are recommended for routine maintenance monitoring. They are presented in two levels of complexity for systems with increasing technical resources. They are further summarized in Table 14 and cross-referenced against well problems in the matrix in the Appendix.

Table 14   Summary of Maintenance Monitoring

Physical inspection

Borehole video

Surface facility inspection

Examination of pulled components

Assessment: Surface and component inspection is essential on a regular basis to judge condition and spot deterioration. Borehole video (color preferred) is a highly useful inspection tool to spot deterioration downhole.

Recommendation: A regular program of inspection should be implemented for all wells. Annual (or regular, more frequent) borehole video inspection is highly recommended to detect and pinpoint problems. When pump repair is necessary, components should be inspected to determine the type of deterioration and photographed.

Hydraulic performance (See methods assessment in Table 8)

Assessment: Necessary to spot changes in performance.

Recommendation: Regular drawdown and flow measurements, schedule determined by site need or sampling schedule. Maintain a maintenance schedule on all instruments and equipment.

Physicochemistry (See also Table 9)

(1) Inorganic parameters

Electronic colorimetric instruments

Spectrophotometers

Electronic pH and mV meters

External laboratory analyses (Level 1 - 2)

Assessment: Electronic colorimeters are sensitive at low constituent levels, and economically provide accurate readings, multiple analyses on single instrument with reagents and interchangeable lamps. Portable and well suited for wellhead analysis of Eh-sensitive parameters.

Spectrometers have significantly higher purchase cost than colorimeter, but considered more accurate than colorimeter, more skill required, multiple analyses on single instrument with multiple lamp range.

pH-mV meters are ubiquitous on environmental projects, and required to perform useful objective analyses of pH and Eh. Require pH-mV instrument and pH and ORP electrodes with reference electrode and solutions. Requires some knowledge of analytical process to operate properly. Eh analyses require flow cell, recommended for pH and conductivity. mV instrument can also accommodate ISE probes for detecting single ion types. Eh should be cross-referenced with redox couple ratios such as Fe^{2+}/Fe^{3+}. Electronics are sensitive.

External laboratory analysis provides the highest analytical accuracy, but potential for alteration of pH- and redox-sensitive parameters, slow reporting, most expensive option for maintenance monitoring.

(2) Suspended particulate matter

Manual volumetric sand/silt collection

Turbidometers

Particle counters

Table 14   (Continued)

Assessment: Manual volumetric methods are suitable for well-testing and diagnosis of sand- or silt-pumping, but awkward in routine maintenance. Turbidometers provide a means of automatically monitoring relative amounts of particulates in pipelines, but give no information on the material. Particle counters are suitable for better distinguishing the volume and size of particles.

Recommendation: Regular schedule of analysis of parameters of maintenance concern, as determined for the site (e.g., Fe, Mn, S, both total and various ionic states), pH, Eh (electrode and redox couples), suspended matter using method best suited. Establish a maintenance and calibration schedule for all instruments.

Biofouling microbial component
(1) Sampling:

Pumped grab methods (Level 1 - 2)

Biofilm collection on surfaces or in filter (Level 1 - 2)

Assessment: Pumped sampling suitable for viable culturing (e.g., BART or equivalent methods, best done in a time-series format), can be done with available sampling gear and containers; high potential for false negatives in sampling for microscopy. Biofilm collection necessary for suitable samples for microscopy (see text), requires collection apparatus (modest expense items). Pumped sampling requires knowledge of collection methods and strategy. Collection methods require knowledge of procedure and care in obtaining samples from the collector.

(2) Analysis:

Light microscopy (Level 1 - 2)

BART methods (viable, heterotrophic aerobic/anaerobic) (Level 1 - 2)

HPC methods (viable, aerobic heterotrophic) (Level 2)

Other cultural methods (Level 2)

Biofilm minerological analysis (Level 2)

Assessment: Light microscopy (moderate expense, skill and experience required) suitable for observing "iron" and "sulfur" bacteria and biofilm solids, but misses single-cell component. BART methods, moderate expense per sample unit, qualitative only, versatile analysis (analysis software available, expert recommended), no laboratory facilities required (see text). HPC methods require laboratory facilities, provide semiquantitative results, versatility in media but representativeness a question. Unlikely to be cost effective in maintenance monitoring for environmental facilities. Other cultural methods (e.g., for *Gallionella*) experimental only. Minerological analyses (XRD, EDS) recommended for selection of rehabilitation chemicals and testing for contaminants adsorbed to Fe and Mn oxides.

Recommendation: A combination of time-series sampling and BART analysis and flowcell/moncell collection and microscopic, minerological analyses provides the most information most cost effectively. Specifics depend on site needs.

Table 14   (Continued)

Record-keeping
Paper files (Level 1)
Database-spreadsheet (Level 1 - 2)
Commercial record-keeping software (Level 2)
   Assessment: Files are indispensable for well maintenance. Paper files
will be needed along with software methods, unless scanning into videotext
records is an option. Database-spreadsheet methods can be done on standard
business software; permits more sophisticated analyses and graphical
display. Commercial software is prepared especially for well-maintenance
applications, is easy to use and graphical for presentations, good for
hydraulic factors, but current products do not factor in biofouling.
   Recommendations: Build a well-maintenance record-keeping system
using existing facility database storage/retrieval and analysis software, plus
hard-copy files. Systems can be customized by knowledgeable experts. All
physicochemical, hydraulic, and biofouling data should be stored. Store
video tapes and photos, as-built drawings, and logs in hard-copy files with
back-up. Back-up everything. Go to specialized well-maintenance software
if it meets site needs.

Note: Levels refer to text recommendations. A diagnostic matrix cross-referencing methods
to problems is included in the Appendix.

Selection of a monitoring program should only be attempted with the assistance of persons very familiar with current methods and their application because
evolving methods may be refining those described in the literature. Actions that
should be taken as a result of detection of deteriorating conditions are discussed
in the next section.
   Level 1 assumes a level of sophistication typical of trained facility operators
who perform routine analyses of water treatment and wastewater plant performance and water quality, and have a modest maintenance budget for a small- to
medium-sized facility. Methods described are expected to be used by these
personnel rather than professional microbiologists and chemists.
   Level 2 recommendations are enhancements that assume the existence of a
sophisticated laboratory program (on-site or available to regular consultant visitors) that is capable of performing limited research. However, some more sophisticated items (e.g., particle counting and automatic data gathering) may be linked
with simpler physicochemical or biological monitoring methods and actually
make the monitoring task more practical.

Level 1 (Operational)

   Level 1 recommendations, as well as those for *Level 2*, are divided into those
for monitoring physicochemical water quality, indications of biofouling, hydraulic performance, and record-keeping. *All of these aspects of monitoring are
important and cannot be neglected.*

Physicochemical Water Quality

Changes in physicochemical parameters, such as increases or decreases in Eh, pH, conductivity, and Mn, S, or Fe concentrations, are indicative of change in the well environment. Increases in turbidity or particle counts indicate increased suspended solids content that may result from silting or fouling. All of these changes can affect the operation of a treatment system and, certainly, its wells. A basic level of diagnosis can be conducted with minimal equipment, as described.

(1) *File review of required monitoring data.* All monitoring and remediation facilities are required to analyze for a range of inorganic and organic parameters that include a number that have possible links to biofouling (N, P, SO_4^{2-}, Fe, Mn, total organic carbon, and individual organic constituents).

This monitoring is conducted relatively infrequently, but can be used as a basis for professional assessment of the likelihood of biological or other chemical fouling and its degree of severity. The problem usually lies in collecting the data available into a form usable for maintenance analysis.

(2) *Regular on-site monitoring.* Monitoring should include: (a) regular file review of required monitoring data, and (b) total Fe and Fe^{2+}, S^{2-}, SO_4^{2-}, Mn, and other relevant parameter measurements by electronic colorimetric or spectrophotometric methods in-house (viable reagents and calibrated instruments), pH, Eh, and temperature by electrode. Eh taken by electrode immersed in an appropriate flowcell should be compared to ion ratios for relevant ion pairs such as Fe^{2+}/Fe^{3+}.

Water quality chemistry changes should be noted and causes determined (seasonal, increased biofouling, arrival of plumes, pulling in deeper groundwater, etc.). The best way to do this is by graphically charting the information and making correlations between data such as redox couple ratios (e.g., FeIII/FeII) and temperature, for example. By eliminating seasonal and daily trends, or external factors such as plume effects, the effects of well deteriorating effects may be revealed in analyses.

Well and Pump Performance

Well performance is an essential part of well maintenance monitoring. All pumping wells should initially be designed or retrofitted for wellhead flow and drawdown monitoring, as well as easy measurement of pump and motor characteristics and pump performance. Borch, Smith, and Noble (1993) make specific recommendations for hydraulic monitoring of wells. Numerous sources, including pump manufacturer literature and the Water Systems Council (1992), make recommendations for pump monitoring. Monitoring should include:

1. File review of past data and interviews with experienced personnel, if any, to determine if there has been a history of performance deterioration.
2. Regular measurement of static water level (in all wells) and drawdown in pumping wells as recommended (may require modification of wellhead) from a consistent, surveyed datum. Alternatively, substitute bailing or slug test data for pumping test data in many monitoring wells.

3. On pumping wells, measurement of flow from a wellhead flowmeter that is itself clean, calibrated, and working properly. Flowmeters are prone to fouling and become inaccurate. Alternatively, for monitoring well pumps, pumping into a calibrated bucket. Measurements of drawdown and flow should be conducted quarterly or monthly, or (as is increasingly the case) may be recorded electronically at more frequent intervals.

4. For pumping wells and well arrays, DWL measurements ("static" and "pumping" levels) taken in monitoring wells or piezometers of known construction, elevation, and distance from the pumping wells can be used to detect changes in well loss. These data are also, of course, used for plotting time- and distance-drawdown curves that provide the basis for detecting changes in aquifer hydrologic properties. Annual step tests of pumping wells against both system pressure and open discharge (as possible in contaminated situations) are useful for providing both pump and well hydraulic curves to detect deterioration in performance (using procedures outlined by Borch, Smith, and Noble, 1993).

5. Pump performance itself is a critical aspect of well performance. In many wells, it is the loss of pump output that spells failure, not excessive clogging of the well screen. Pumps can be monitored in place in several ways, among which are: motor power draw readouts (kWh, amperage draw) with manual instruments that can be charted over time or compared to manufacturer data and, of course, wellhead flow. Pumps that are readily removed and disassembled may be pulled and inspected on a regular basis (e.g., annually) for wear, corrosion, or clogging.

Biofouling

Biofouling analyses should be conducted for both viable (culturable) microflora and nonculturable but identifiable microflora such as filamentous or stalked "iron bacteria." Analyses should include:

1. *Examination of slides.* Exposed in the well or well outflow for a period of usually not less than 1 week (adjusted to meet local needs) via light microscopy (400X to 1000X magnification) for signs of filamentous and/or stalked bacteria associated with Fe-precipitation.

2. *Use of prepared biofouling semiquantitative methods (e.g., BART kits) inoculated with pumped samples.* Pumping should be conducted after a nonpumping period (not less than 2 h) and proceed for not less than 30 min. Several BART tubes should be exposed over the pumping period. BARTs may be exposed on consecutive days to check for the effects of differential sloughing on reaction times (d.d. data). Cullimore (1993), which is the "book" on BART use, recommends several time-series sampling procedures, for example, scheduling a well shutdown,

sampling before the shutdown, then sampling on an essentially log-time schedule through a day.

Alternative for biofouling monitoring step (1). Microscopic examination of sand from aseptic, prepared moncells or other filter devices, using the methods of Howsam and Tyrrel (1989).

Alternative for biofouling monitoring step (2). Use only in labs with a competent graduate-level microbiologist familiar with the methods, equipped for preparation of and handling of cultures for microbiological testing ("Don't try this at home, kids!" — these methods are more at home in Level 2, discussed in the following sections):

R2A heterotrophic plate count medium may be used to replace SLYM-BART) since it supports slime-forming pseudomonads pretty well (See APHA, AWWA, WEF, 1992). However, to enrich or isolate for pseudomonads or sulfur-oxidizing slime-formers, consult, e.g., Cote and Gherna (1994) and Holt and Krieg (1994) or Atlas (1993).

W-R (see Cullimore and McCann, 1977, or Cullimore, 1993), or R2A+FAC (R2A amended with ferric ammonium citrate; Smith, 1992) provide organic-Fe complexes and serve the same function in a plate environment as IRB-BART.

Culture SRB in Postgate-type roll tubes; see Holt and Krieg (1994) or Atlas (1993).

Note: Agar plate methods such as HPC media and "generic" broth media such as those for SRBs do not employ precisely the same media formulations nor provide the same growth environments as the BART tubes, and potentially recover a different array of microflora.

Records

Measurements are useless without records to provide a history of measurements over time. Paper files and manual analysis are entirely suitable; however, most environmental monitoring and extraction well arrays are managed by operations that are computerized. Maintenance data recording and analysis is possible using either commercial business database-spreadsheet software or product-specific or specialized software.

A software system designed specifically for well maintenance management can be used without modification to record and display graphically at least most of the data helpful in well maintenance analysis. Drawbacks to the available software are: (1) the money devoted to purchase of software for a single purpose and (2) data analysis limited mostly to changes in hydraulic performance. In remediation extraction wells, once performance declines, preventive maintenance is no longer possible.

General interactive database-spreadsheet software has the disadvantage out of the box of not being designed for your purpose (the programmers never heard of well maintenance). It does have the elements you need: the ability to store data in formats you design, permits analysis of data, and graphical displays of existing data, changes over time, and the ability to make projections. A qualified programmer who knows what you want can make the necessary modifications to provide the well record-keeping and analysis system that meets your specific needs and system capacities.

An example of when computer-aided analysis is especially useful is making sense of BART reaction-type and day-of-delay (d.d.) values. The trends they reveal are best analyzed graphically by looking at the microbial ecology "signatures" of well environments and changes in d.d. values over time.

These analyses can be made using the commercial BARTSOFT™ software being developed by the BART system developer (Cullimore, 1993) or by database spreadsheet. If there is a change from the previous sampling period, repeat the test.

If you did not necessarily understand this example, realize that this kind of analysis requires some direct training with biofouling analysis-experienced people reviewing your data. This is an example of the qualitative state of the biofouling analysis art at the present time, but also that graphical software takes much of the mystery out of analyzing trends in indicators of possible well deterioration.

Level 2 (Enhanced)

Physicochemical Water Quality

Certified laboratory instrument-grade methods are preferred by regulators and courts for inorganic parameters measured to take regulatory action. These can provide increased sensitivity, but it is important that transport time for redox-sensitive samples is very short. In addition, the following are recommended to better characterize the organic component:

1. Add to Level 1 organic carbon parameters, N, P, and nutrient assessment, and Fe^{2+}/Fe^{3+} and dissolved oxygen for changes in redox conditions pointing to increased "eutrophication." Such analyses would probably be run in any case on an *in situ* bioremediation site. Examinations of changes in anthropomorphic organics present can indicate biotransformation and concentration changes indicative of microbial sorption.
2. Add particle counting or turbidity measurements to detect changes or increases in particulates. Particle counting can be a useful means of detecting a fouling situation in development (Hargesheimer, Lewis, and Yentsch, 1992; Cullimore, 1993) or detecting filter breakthrough. These are especially useful if they are automatic. A history of particle counts must be gathered in order to make useful interpretations.
3. Add elemental and X-ray diffraction analyses of biofilm materials for selection of treatment chemicals and detection of possible metal contaminants adsorbed to biofilms. These methods are relatively inexpensive and available in many larger communities or university laboratories.

Well and Pump Performance

Primarily, at Level 2, continuous automatic data recording methods should replace any manual measurement and recording, as long as the automatic system can be considered reliable. There should be a capability to analyze data vs. changing pumping conditions in the wellfield (various pumps on or off, regional hydrologic changes).

Automatic equipment is vulnerable to environmental attack and can become inaccurate. As well, the operation and condition of the automatic equipment should be regularly reviewed as part of the maintenance program.

Biofouling

Despite the limited information available from microscopic examination, Level 2 systems may wish to add photographic recording of biofilm samples and scanning electron microscopy for exposed coupons of known composition (e.g., discharge pipe metal or plastic) to quantify corrosion or other deterioration over time. Even if there is no attempt to quantify biofilm samples, photomicroscopy is recommended to provide a visual record of distinctive biofouling organisms and mineral accumulations.

The monitoring program should institute a side-by-side comparison test program of two microflora detection methods such as the BART system and Standard Methods, W-R, or R2A+FAC HPC agar to provide greater information on the full microflora present. This information is still site specific and not transferable between locations. It has to be "calibrated" for each well system.

Media comparisons should be made under the supervision of a microbiologist familiar with specific biofouling detection methods. Over time, such microbiological analyses should become more statistically reliable as monitoring programs develop a history.

Records

There is not a great qualitative leap to be recommended over Level 1 except for increased sophistication of statistical comparisons of data. The nature of the statistical program is site specific and can be decided on a case-by-case basis.

Frequency of Monitoring

Current research on well maintenance monitoring, including biofouling (Smith, 1992; Cullimore, 1993; Borch, Smith, and Noble, 1993) does not provide a definite answer to the questions of how often and how long monitoring is necessary. The wells studied in the research projects cited by Smith (1992) and Cullimore (1993) were already biofouled and provided positive but variable results in analyses and much variability in the symptoms of biofouling. Deliberate proactive well maintenance planning is only now beginning, and the history is not written yet.

The optimal approach in designing a monitoring plan still is to field test a preliminary program at the wellfield of interest.

1. Wells could be sampled and analyzed weekly for a period of time to provide a baseline of information about water quality and hydraulic parameters. Frequency of analysis can then be reduced as the situation warrants.
2. Many facilities may decide to sample in conjunction with their regular regulatory monitoring of chemical quality. This approach is acceptable if the frequency of monitoring is adequate to provide the information needed to assess well deterioration. For example, quarterly sampling is probably the minimum frequency for early-warning detection of biofouling deterioration.

RATIONALE AND COMMENTARY

While not ultimately conclusive or a precise step-by-step recipe for maintenance monitoring well deteriorating conditions, methods such as those presented here can provide sufficient early warning of deterioration to allow for effective control if employed properly and used faithfully.

Level 1 procedures provide the type of information suitable for use in making an informed professional judgment as to whether or not a well is deteriorating, and what the nature of that deterioration might be. With such measurements over time, it might be possible to make an assessment of relative deterioration or improvement. *These procedures, first developed for water supply plant operators, are being modified for and are practical for most environmental facility operators and do not interfere with normal operations.*

There is not enough information on biofouling monitoring in practice to conclusively justify the added expense of specialized biofouling and chemical analytical procedures under Level 2 in routine operations at this time, but Level 2 automation and particle and hydraulic data gathering and analysis have engineering benefits and save on labor costs. They are being implemented separately from well maintenance considerations for these reasons.

In actual practice, a customized monitoring program may employ methods at several levels of sophistication. For example, a system that may choose a Level 1 biofouling monitoring approach may have already been fully automated for hydraulic flow measurements (Level 2) and have someone on staff interested in running detailed statistical analyses (Level 2), but may not have access to automatic drawdown measurement equipment, necessitating manual (Level 1) drawdown measurements.

At any sophistication level, there still is no direct correlation between microbiological data (such as BART, CFU/ml, or slide results) or particle/turbidity data to terms such as "lightly biofouled" or "heavily biofouled." At the present time, judgments as to whether a well is more or less biofouled than before can only be made on the basis of monitoring over time.

In general, maintenance monitoring approaches should be tried and reviewed over a period of time and adjusted based on experience. They must be implemented as part of a systematic maintenance program involving:

1. Institutional commitment to a goal of deterioration prevention.
2. Maintenance monitoring as part of an overall site and well system maintenance program.
3. A method evaluation to improve performance.

Maintenance monitoring has to be performed to prevent severe eventual deterioration in environmental well systems. However, it has to be recognized that monitoring approaches and responses will be site specific and likely will require adjustment during implementation.

7   Preventive Treatments and Actions

Maintenance monitoring does no good if no action is taken to control deteriorating effects once they are detected. Likewise, waiting until performance deteriorates markedly raises the odds that treatments are less than totally effective. Preventive treatments represent a "proactive" treatment approach, to borrow a term from current pop-psych lingo. A preventive treatment is that which is applied before performance (however defined) declines.

If well performance-degrading problems such as sandpumping, corrosion, or biofouling are detected by monitoring, then what? At this point, preventive maintenance repairs and treatment are implemented. The nature of the treatment depends on the nature of the problem.

Preventive treatments must be implemented upon detection of the indicator signs of a problem and before performance deteriorates. To do this effectively, well systems must be equipped for preventive monitoring and treatment (Chapters 4 and 5) and a maintenance program must be in place (Chapters 5 and 6).

SAND/SEDIMENT PUMPING

Sedimentation in a monitoring or pumping well can possibly be controlled by redevelopment using methods described earlier (Chapter 4). Methods chosen should be appropriate for the screen and aquifer. An attempt should be made to develop the well until it has a very low level of sediment in samples (<50 mg/l). In some wells, especially monitoring wells in formations with layers of fine material, this may not be entirely possible. Sediment may have to be minimized by purging during sampling events. Surging and jetting should only be performed by well rehabilitation contractors or crews experienced with these methods.

In pumping wells, sediment production can be greatly limited by installing an SFCD, which forces a cylindrical well in flow as described earlier. This reduces or eliminates turbulent upflow through packs that are too open, and limits or eliminates selective channeling through the pack in the top 15% of the screen. The SFCD can be installed as original equipment in a new well, or retrofitted.

Chronically sandpumping wells that cannot be repaired in any other way or replaced should be fitted with sand removal devices. Probably the most useful is the in-well centrifugal desander, such as the Laval model. This desander fits over the pump intake (Figure 17). This protects the pump and system.

Sediment will be dropped to the bottom of the well and may have to be cleaned out on a regular basis. Some limited research (Lehr, 1985) indicates that a desander will force a new equilibrium in the well over time, eventually reducing or halting sand production. As a last temporary resort until something else can be done, sand-resistant pumps should be employed. Such pumps should not be considered a permanent solution to a sandpumping problem.

Sanding/sedimenting monitoring wells are very difficult to fix because of their already restricted diameters, which precludes rescreening or SFCD installation. Redevelopment may work, using small-diameter surging tools or (as a last resort) overpumping. If not, purging should proceed before sampling until the water clears or, alternatively, samples should be filtered. The risk of relying on sample purging and filtration is that sorbed constituents may be lost to analysis, or at least difficult to quantify.

As the situation permits (budgets, time, permission), sanding monitoring wells that interfere with sample quality should be replaced with better designed, constructed, and developed wells.

CONTROLLING CORROSION

The primary method of controlling corrosion is to design properly the well and treatment system to prevent corrosion. Water quality information should be used before design decisions are made. If corrosive conditions exist, proper casing and screen materials can be chosen. Stainless steel and plastics (as normally used in monitoring and remediation systems) are resistant to most (but probably not all) corrosion caused by underground environments and chemicals added during rehabilitation.

In corrosive environments, stainless steel and plastics are preferred to low-carbon steel, and their specification assures the well operator long-term resistance to corrosion under most conditions. It should be remembered though that stainless steel does corrode (especially screens, in which the metal is stressed and altered by welding). The stainless alloy process only slows and does not stop microbial-induced corrosion. In addition, it is important to remember that not all stainless steel is the same (not even within types), and their characteristics should be matched to the well environment (Chapter 4).

Where external sources of corrosion, such as high-voltage power lines, are detected, cathodic protection of steel casings may be specified on certain systems. Experience has shown that where microbial activity is generally absent, protection of steel is achieved when the potential is depressed to −850 mV; but where microbial activity is intense (or likely to be, the most normal case), the potential has to be depressed to at least −1000 mV. This voltage drop may be extremely difficult to achieve over the length of wells in many situations and may effect water quality near the well.

BIOFOULING

If caught early, regular chlorination and stepped-up vigilance is recommended to keep biofouling problems under control in many pumping water supply wells. However, chlorination usually is not an option in monitoring wells or remediation wells due to the need to control purge fluids and chlorine's temporary effects on water quality around the well.

In many cases, real well plugging does not occur, only maddening and irregular poor water quality problems. For these situations, periodic treatment with a nonoxidizing chemical such as an organic acid (e.g., acetic acid), combined with surging, can be beneficial in keeping biofouling below problem levels. An example of an acetic acid product is LBA (CETCO Inc.). LBA is listed as conforming with NSF Standard 60, which is convenient in documenting and justifying well treatment choices.

Preventive Treatment Chemicals

Acetic acids are liquids and quite potent in industrial strength solutions (70% acid). (This is not your mother's vinegar.) The acetic solution chosen should be as free as possible of impurities introduced during manufacture.

Sulfamic acid is a solid acid form and also degrades biofilms. It is stable and relatively safe to handle and mix, and also causes minimal problems with organically laced purge water. It should also be highly pure and inhibitors used around rubber and metals for safety's sake. One problem with some sulfamic acid is ammonia formation. For this reason, sulfamic acids may not be approved by authorities on certain environmental sites.

Peroxide-based solutions provide the oxidizing and disinfecting effects of chlorine compounds without the complications of chlorinating organics and making them more recalcitrant and toxic. Peroxide itself rapidly degrades to water and oxygen and dissipates harmlessly. Such powerful oxidants do severely interfere with aquifer water quality at the well, at least for a short time. Peroxide solutions in general (50–75% H_2O_2) are hazardous to handle (this is not the stuff in the medicine cabinet) and should be kept clear of chemicals that react with oxidants.

Augmentation of Chemical Treatment

Other possibilities in maintenance treatment exist; the most practical is heating chemicals to boost their effectiveness and to assist in killing and dispersing biofilms. This is the principle behind the Blended Chemical Heat Treatment (BCHT) described in the following under Rehabilitation (Section 2.2).

"Pasteurization" alone has limited application, as demonstrated by field tests. While effective in practice and lacking chemical reactivity problems, hot water recirculation (the most practical method) (Figure 27) requires a very large heat input and is relatively slow (Cullimore, 1981, 1993). Heat is also cumulative: most

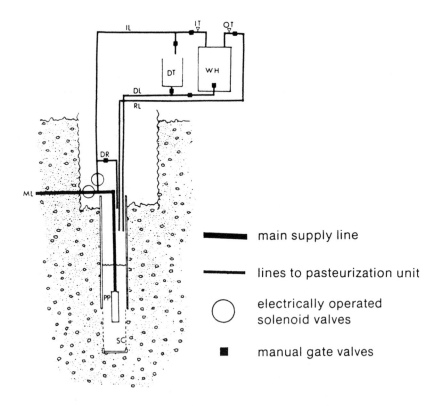

Figure 27.   Hot water recirculation system for wells: DL, drain line to heater and drum; DR, drain for emptying IL; DT, 45-gal drum for flow test; IL, intake line; IT, intake thermister; ML, main line to treatment plant; OT, output thermister; PP, pump; SC, screen; WH, water heater. (Cullimore, 1981.)

aquifer materials subject to regular heating store heat, and this can lead to drying and cracking of grout and surface structures. Excessive heat is also not favorable for most plastic components used in wells, but temperatures applied (<60°C in the well) are within the tolerance of PVC and FRP casing pipe. If misapplied, heating can actually encourage growth at the edge of the thermal impact zone.

Radiation has also been successfully tested as a biofouling limiting agent. A U.S. patent exists (#4,958,683, William Rogers and George Alford) for an in-well cobalt-60 gamma-source tool. Gamma irradiation serves to generate hydrogen peroxide *in situ*, which serves as the biofouling degrading agent. Based on experience, interest in such a tool is likely to be limited despite its promise due to regulatory and personal constraints.

Biofouling Recurrence

Whatever the treatment, biofouling always grows back — the trick is to treat it again as early as possible. Monitor to head off a comeback using methods described (Chapter 6). If regrowth persists, look into regular or continuous chemical treatment. A maintenance contract between a competent well cleaning contractor (qualified for the site conditions in question) and the well system operator is a good idea.

INORGANIC ENCRUSTATIONS

Preventive chemical encrustant removal is specified in wells that have encrusting waters, where mineral salts tend to build up on slots. These can be removed by appropriate acidizing or by sonic vibratory treatments as described elsewhere (e.g., Borch, Smith, and Noble, 1993). Such encrustation often occurs along with biofouling and must be removed to permit effective biofouling treatment.

COSTS AND TIME OF ROUTINE PREVENTIVE MEASURES

Costs of well deterioration impacts were discussed previously in Chapter 2. Maintenance measures have costs associated with them, too. However, the concept of maintenance is to incur the costs of operation here as a regular investment in preserving the assets vs. the uncontrolled depreciation of facility (well) deterioration. The investment in maintenance monitoring and maintenance preventive treatment should be recouped through reduced operational problems, energy and treatment costs, and the reduction or elimination of rehabilitation and its interruptions and emergency costs. The method used by Sutherland, Howsam, and Morris (1994), described in the following, can be used for these analyses.

Maintenance Cost-Benefit Analysis

This cost and time investment should be compared to the available history of equipment renewal, increased pumping costs, well rehabilitation, and other maintenance costs such as line flushing and chemical costs that are impacted by fouling and corrosion originating in the well. Valuable history may come from the past at this particular facility or experience elsewhere that applies.

Cost-Benefit Analysis: A Spreadsheet Approach

The report to the British Overseas Development Agency by Sutherland, Howsam, and Morris (1994) provides a rational basis for cost-effectiveness comparisons of operating wells with and without monitoring and maintenance. While developed principally for water supply well projects, the methodology is directly applicable to environmental well system management.

The Sutherland et al. system is based on a spreadsheet analysis ("what-if") approach that takes into consideration a variety of levels of:

1. Maintenance monitoring (and associated costs).
2. Maintenance action levels.

3. Well classifications based on historical assessment of well deterioration in a variety of hydrogeologic settings.
4. Costs and intervals of rehabilitation with and without maintenance actions.
5. Costs of operation with and without maintenance actions.
6. Methods of factoring in "intangible" external factors in making decisions about well maintenance, such as the existence of high standards of product water quality, the need for uninterrupted water supply, and regulatory considerations (consequences of poor water quality or supply interruption).

The system acts as a convenient means of encouraging a site manager to get and organize relevant well operational information, but also allows the maintenance analyst to guess about costs and intervals to some degree.

Weighting factors are provided to quantify the relative importance (high — medium — low) of:

1. "Societal factors" (read regulatory official trouble in the case of unreliable environmental wells).
2. Availability of alternative supplies (assume irreplaceable for environmental well arrays).
3. "Environmental factors" (see item 1, plus consider environmental impacts on well performance).

The Heartbreak of Well Failure: An Overriding Weighting Factor

Environmental well systems, especially those for remediation pumping, have historically been high maintenance because they are at risk for costly severe deterioration (see Chapters 1 and 2). Consequently, for facility managers, if maintenance practices save money and keep the system working over the life of a facility (and this assertion can be quantified using methods such as that described above), maintenance is cost effective.

However, costs are not insignificant. At present, a maintenance monitoring and treatment program can justifiably cost (1) 100% or more of projected rehabilitation and reconstruction costs or (2) the annual amortization of the well array over the project life.

Why do it then? A couple of reasons come to mind:

1. The money is spent in a controlled manner (whatever the costs happen to be).
2. The system remains operational and providing the required results (pretty much regardless of cost).

The alternatives, such as continual well and system rehabilitation, loss of control of a plume, or surprise arrival of a plume at a wellfield, are by comparison infinitely more expensive, making conventional accounting cost-benefit analyses virtually irrelevant.

Maintenance planning and execution have the following projected results:

1. *Buys time, more likely to work.* Although monitoring and treatment cannot prevent well deterioration by themselves, early detection of problems permits effective regular preventive treatments (if needed). Preventive treatments limit the need for extensive and less-sure rehabilitation.
2. *Budgeting and use of resources.* The maintenance cost is spread evenly over time instead of in episodes of "emergency" expenditures. It is also accounted for in the normal budget and therefore funded. When pumps have to be pulled for proactive maintenance repair, the required work is accomplished more quickly and the damage is less severe.
3. *Scheduling.* Service interruptions are eliminated or less frequent and capable of being scheduled. Wells may be occasionally off-line for preventive treatment, but can be brought on-line quickly in case of need.
4. *Better downstream results.* Allied problems (biofouling and discoloration in distribution lines, water treatment costs, filter backwash intervals) are brought under greater control.
5. *Longer facility service life.* Any time added beyond normal depreciation is a financial plus in the life of such systems. In addition to accounting depreciation, time expenditures for design and approval of replacement systems is delayed, and construction of new systems is also delayed.

What are these costs then? They are highly variable and difficult to compare among situations. Here are some example values, which readers should check for in their specific situation.

Costs of Maintenance Monitoring

Costs of equipment and materials are quite subjective. Equipment and instruments recommended can be purchased through laboratory and environmental supply sources, or can be fabricated under license or based on public-domain designs.

Water quality. The total cost for the first year of monitoring pumping wells, assuming there is no laboratory equipment at all, purchase of quality field electronic instruments, and using BART kits, with a monthly sampling program under Level 1 protocols, would be about $2000 for the first well and about $650 per well for additional BART tubes and other supplies and reagents for additional wells for the biofouling and physicochemical water quality monitoring.

Hydraulic performance. Hydraulic performance monitoring is equipment intensive but requires very little personnel time. Assessing changes in performance is not possible without both flow and drawdown data, and both have to be measured in the wellfield.

Good combination ammeter-voltmeter-ohmmeters such as Amprobes can be found anywhere for less than $100 and should be available to all well operation crews, who should know how to take control and lead readings.

Turbine flowmeter installations may cost up to several hundred dollars each and require wells to be taken out of service for installation. Venturi-type flowmeters cost somewhat less and have no moving parts. Both types of meters themselves must be maintained to be reliable.

Such flow monitoring is normally required in any case for control of the treatment stream as well as regulatory records purposes. Flow monitoring is also absolutely essential to allow the detection of very fine changes in performance before pumps clog or corrode.

Alternatives to permanently installed flowmeters include portable sonic flowmeters. These cost over $1000 each, but may be used on multiple wells and are quite practical for remediation pumping streams with their high particle counts.

Drawdown is conveniently measured by the ubiquitous electric water level sounder ($300–600), air line (capital expense plus air source <$200), transducer-based systems, or sonic water level sounder (each >$900).

Sonic sounders and air lines allow drawdown measurements access in wells with limited or no casing access, but these instruments have an accuracy of approximately ±6 in. (152 mm) and ±1 in. (25 mm), respectively, vs. approximately ±0.03 in. (0.76 mm) for electric water level sounders. In addition, air lines clog under most remediation well conditions and have to be properly maintained.

Transducer-based water level sensors provide accuracy, low maintenance, and easy adaptability to automation and data recording. These are preferable for arrays of pumping wells for routine maintenance.

Adding flowmeters and additional instrumentation (automatic water level recording, motor controls such as Franklin Subtrols) to pumping wells may boost well costs by several $100s to over $1000 per well, which is recouped in reduced future motor burnout, rehabilitation costs ($5000 per well per incident instead of $2000 once + time and minor equipment maintenance), and intangible costs of having critical wells performing badly.

Actually, all or most of these devices are normally already present on the site for other testing needs and do not represent new costs dedicated to maintenance. Maintenance is just one more good reason to install, maintain, and faithfully use them.

Personnel time. One hour per well per month of operator time is a reasonable assumption using manual methods for a full suite of Level 2 analyses (more if apparatus maintenance is required). This can be reduced and streamlined if particle counting and hydraulic parameter measurements are automated. Sampling intervals may be more or less frequent and perhaps focused on specific (troublesome) wells.

There may be an additional cost of professional assistance, and again, the costs are variable and subjective, but probably worthwhile in getting off to a good start.

Preventive Treatment Costs

The treatment costs themselves, if an assessment of monitoring data calls for such treatment, are highly variable. They can probably be most closely correlated to personnel time involved. The chemicals and other materials themselves are relatively inexpensive. What is expensive is the time of skilled well cleaning crews and their equipment, comparable to drilling crews (>$1000 day plus materials) for maintenance work. These kinds of costs may mushroom, of course, when there are additional requirements such as personal protection and highly contained handling of purge water.

A goal in cost projection, as mentioned previously, should be to keep life cycle maintenance costs within projected rehabilitation costs, and annual costs within the 10-year amortization figure for the specific installation. This should be split between monitoring and treatment.

Improving Cost-Effectiveness in Maintenance

While it is justifiable for maintenance costs to equal 10-year replacement costs, it is not necessary. There are ways to keep costs down without compromising the maintenance program.

1. *Design for maintenance.* The first step to make the best use of maintenance funds is to design well systems so that they are readily and easily treated as necessary (Chapter 4). A well-designed system:
   a. Resists deterioration and does not need as much maintenance.
   b. Wells and treatment systems should be designed for rapid and easy monitoring with automated hydrologic instruments and convenient sampling.
   c. There should be good access, easy hookups, readily available hauling service, and other steps taken to minimize the fuss and labor costs associated with maintenance and rehabilitation treatments.
2. *Diligence in maintenance.* Keep the well-designed equipment in good shape:
   a. Monitoring has to be diligent to detect problems while they are minor. Good design facilitates this.
   b. Have spare pumps and replacement parts readily available and have personnel trained in the diagnosis and repair of pumps. Contract pump service contractors should be on-call if on-site personnel are the first called on for service.
   c. Apply preventive maintenance treatments as soon as detectable problems occur. Good design, access, and planning help to make this more likely.
   d. Check results and refine the monitoring and treatment based on experience, and incorporate new methods and information to fine-tune your system.
3. *Negotiation.* This is business and the site management is buying a service. There is competition and the service and equipment providers in the business want the projects. If in checking around, the site manager finds that costs for maintenance services he or she is using are out of line, it is time to find out why and make adjustments as necessary.

It may be that the services being provided are indeed superior (usually a temporary situation) or that the conditions for maintenance on the site result in higher costs (fluid concentrations, rapid deterioration, and frequent maintenance). On the other hand, the site manager may be getting soaked on chemical or instrument costs. Compare costs for comparable performance, but definitely compare.

Sidebar: *WARNING: Dealing with Well Cleaning Chemicals*

Well treatment chemical represent serious chemical hazards if mishandled. The typical well treatment chemicals are powerful oxidants and acids. Both can burn people or react with chemicals or equipment on the location. The site health and safety officer should be aware of the well treatment procedures and chemicals proposed for use, and should advise of any possible problems. In general, personal protection is the responsibility of the well treatment service provider. They should have a safety response plan and make it available to the site safety officer.

Well treatment chemicals should only be selected and their doses and frequency of application determined by professionals well versed in the selection and use of well cleaning chemicals. Such people preferably are in the regular practice of specifying and applying well treatments, and are knowledgeable about environmental site limitations, and treatment chemical hazards.

The most common mistake in well chemical application is selection based on assumptions about problems and limited information on chemicals, derived from product literature and old papers, or other semireliable information. *There are no reliable recipe charts meant for self-selection of chemicals.* Many problems have been made worse by application of the wrong chemicals in the wrong doses, usually compacting the fouling, or driving it out of the gravel pack and onto the borehole wall, or enhancing microbial growth.

Chemicals and doses should be based on a thorough analysis of the nature of the fouling problem as previously described. Common mistakes are (1) too little oxidant and (2) too much acid, or use of oxidant where application of chelating acid may be more appropriate. This is where monitoring really pays for itself: in precise application of preventive treatments. Maintenance monitoring methods as described in Chapter 6 are also useful in a diagnosis mode where preventive monitoring has not been practiced.

All such powerful cleaning chemicals should only be handled and applied by professional well-cleaning crews trained in their use. Site management without specific experience in application of these chemicals should not attempt to train site maintenance personnel without additional assistance. If experienced treatment contractors are requested to train site personnel to apply chemicals for maintenance, they should make sure that the facility's crews are fully proficient and provided with all the necessary information and equipment (mixing, application, safety, and disposal) to perform the job. A formal training program with testing of comprehension may need to be developed to assure proper training and retention of the training information. *Facility managers and consultants absolutely should not try to self-medicate based on paper research, or limited observations of well rehabilitation contractors alone.*

In general, chemicals are added at the wellhead and directed to the static water level via a pipe. Lighter, less aggressive acids and oxidants should be recirculated through the well pump and throughout the water-filled section of the well casing. Heavy acids should be added at the static water level and allowed to sink.

If possible, chemicals should also be agitated out into the surrounding aquifer. A summary of chemicals and methods is provided in Chapter 9 and is further elaborated by Borch, Smith, and Noble (1993).

III. Rehabilitation and Reconstruction

This is probably where you enter this discussion if you are a typical new and unwilling student of well maintenance and rehabilitation. Here are considerations for well rehabilitation and reconstruction. Once you get it under control, or if you start over, go back and review the sections on prevention and maintenance.

8 Decisions on Rehabilitation Methods: After Things Go Wrong

Performance is already down at the well array: drawdowns are increasing, discharge lines are partially clogged, and filters are backwashing constantly. Now you have to save the system.

Ideally, rehabilitation or well restoration may never be necessary if effective preventive maintenance measures are implemented, or at least only necessary after a long time. The need for rehabilitation is often, however, the starting point of thinking about well maintenance.

Rehabilitation choices depend on the problem. Treatments per se are only used for biofouling and encrustation, where something is removed or suppressed. Preventing or rectifying sandpumping and corrosion require some reconstruction strategy to keep the abrasives out and limit deterioration of the screen and casing, as well as the pump. Sand- or silt-pumping additionally can be addressed through redevelopment. In well rehabilitation, these are essentially preventive activities (materials selection, design, and development) applied reactively.

This and the next chapter cover the issues, recommendations, and techniques for well rehabilitation. Chapter 10 looks backward to case history experience and forward to what needs to be done to progress. Summary matrices cross-referencing strategies and methods to well problems are found in the Appendix.

MANAGEMENT AND SAFETY IN WELL REHABILITATION

Management and safety considerations are mentioned first because any well rehabilitation has to be planned first. Considerations for site management or consultants contracting well rehabilitation services are somewhat different from those for the well rehabilitation contractors themselves. They will be considered here in sequence.

Site Management Considerations

Management's first goal is to get the wells and associated systems up and running in the least time possible. Cost is usually a secondary, but important, consideration, often offset against the costs of losing a well array or failing to meet a clean-up schedule.

Management also has to be concerned with safety and liability problems in well rehabilitation, which have the potential for being the least well-controlled activity on the site. Management also has to be concerned about regulatory aspects of the well treatment.

Responsibility for the Work

One important consideration at this stage (the system has deteriorated until something had to be done) is responsibility for actions and costs. The need for rehabilitation of wells and systems may already have resulted from deficiencies in design, construction, and operation of the soil or aquifer remediation system.

What's important? One result of this realization is that finger-pointing and acrimony may result among site management and consultants, who may have designed and implemented the systems now failing. Regulatory personnel will likely want a review of the causes of the problems and plans for treatment. They may want to exercise review or veto power over potential treatment approaches.

This can result in delays in implementing well rehabilitation while lines of responsibility are determined and requirements satisfied, or worse yet, legal proceedings are instituted. Meanwhile, the systems do not function properly and the purpose of the installation — plume control, recovery, or monitoring — is not fulfilled.

This is a situation in which management considerations rather than technical constraints become paramount. Therefore, management has to keep sight of its mission: the management of a monitoring or remediation array to make sure it can perform its task. Therefore, the goal in any management negotiations should be *to safely and effectively restore the systems to acceptable performance and prevent the problems from being so bad again.*

Good excuses for lack of suitable maintenance. From a strictly technical standpoint, there is not much of an objective basis in such negotiations for establishing anything like fault or liability under some concept such as "due diligence" *because there are no standards for preventive maintenance of such systems. Such preventive maintenance programs are site specific and continually developing.* This publication and the few covering some of the same topic areas since 1990 represent most of the body of evidence. However, standard guides for monitoring well maintenance are being developed, which will certainly be applied to extraction wells to some extent.

Here also is a suitable time to take note that *such environmental control systems are relatively new, and problems with them have not been well known until recently.* Referring back to descriptions of causes earlier in this book, it is a rare product of our educational systems who is conversant with the interactions of all the variety of microbial, hydraulic, chemical, and operational impacts faced by a recovery or monitoring well. Therefore, the consequences of (at the time) reasonable decisions and actions may not have been foreseen, sort of like the result of decisions by another well-meaning manager, Pontius Pilate. *Anyway, the time for excuses is over.*

Getting the Job Done

The pragmatic solution in any such situation is to avoid the laying of blame and to get the job done (but learning a lesson for next time). Environmental site well and pumping service is usually included in the service contracts between site management and consultant engineers. They may choose to subcontract such work, and often do, except where consultant management philosophy is to provide all services from "in-house."

Some companies, for management reasons, then attempt to have all necessary skills from botanists to pump technicians under the corporate roof. This may or may not be a worthy goal from a cost-effectiveness standpoint.

Do well rehabilitation in-house? Maintaining dedicated in-house well rehabilitation teams is a particularly unproductive strategy in planning for well rehabilitation services. One reason is that highly experienced, independent companies already exist that do such rehabilitation work frequently. They are qualified, know the necessary methods, and fully understand the safety considerations involved. When they are done, the work is done and the highly experienced and expensive crew is not on the site management's or consultant's overhead. Keeping such crews on overhead would only make sense if the company (1) uses them frequently and (2) can market such a service to the outside.

Specification writing. Assuming reasonably then that work is contracted out, a second consideration is writing specifications for well and pump rehabilitation and contracting to get the work done. Specifications need to be goal- and not process-oriented. They also need to be based on analysis of the actual problem and the actual operational challenges involved.

Unfortunately, two counterproductive tendencies prevail in specification writing for well rehabilitation:

1. *By the book:* One tendency in well rehabilitation specification writing is to assemble boiler-plate specifications based on a variety of historical documents developed in the water well field. Another tendency is for the specification writer to narrowly define the rehabilitation methods. These are discussed in more detail in the following material on Specifications for Rehabilitation.

2. *Getting off the subject:* Another specification problem is misalignment of the focus of the work. On environmental well rehabilitation projects, this usually takes the form of excessive attention devoted to chemical safety and hygiene considerations. *Obviously, with human life involved, under our prevailing value systems, these are paramount concerns; but specification writers have to have some faith in the experienced con-*

tractors they hire. If reputable experienced contractors are retained, they can be counted upon to work safely while being productive as well.

Solution: *Specifications should be written by people experienced with environmental well rehabilitation and the work performed by or under the supervision of experienced contractors or crews.*

Safety and Productivity in Well Rehabilitation Work

Safety assurance. Safety is a further consideration and, from a liability standpoint, the greatest of all. All site managers in the U.S. are, of course, required to have a site health and safety plan under OSHA regulations (29 CFR Part 1910) and similar requirements exist (or should exist) elsewhere. The site safety plan describes the hazards involved on the site (chemicals, noise, radiation, etc.), levels of personal protection and monitoring, traffic and site-access control, and policies and methods available for decontamination, confined-space entry, and emergency response. This should be supplied to the potential contractors and can be specified as being confidential.

Management should qualify potential contractors both from their technical qualifications and their safety preparedness.

1. All personnel should be OSHA 40-h trained at the minimum, with necessary supervisory qualifications and proof of attending annual refresher courses.
2. The potential contractor should also have a general safety and hazards notification plan to protect their own workers, and include decontamination procedures (routine and emergency) for personnel and equipment.
3. The contractor should be prepared to verify that contractor personnel on-site know about potential site hazards (if any) and how well rehabilitation activities may interact with these hazards.

The main problems are personnel contact with potentially contaminated groundwater and the interactions of groundwater constituents with chemicals that may be used in well rehabilitation. The contractor should be aware of chemical mixing problems and be prepared to use methods that minimize the generation of hazardous purge water or splash. At the same time, site operators or their consultants should have a plan for containment and disposition of any purged material, as well as expendable equipment and supplies such as pipe.

Management should make sure that contractor personnel actually have and know how to use personal protective equipment necessary for the site (even if it is only Level D), air monitoring instruments appropriate for the potential hazards, and means of communication used locally. Personnel should be equipped with any special gear that the site specifies for visitors, such as neon orange vests with polka dots or special-issue radios.

Facilitating productivity. The well rehabilitation or pump service activities themselves have to be worked into the flow of site operations. Personnel need to be briefed on the specifics of the current site hazards (if any) and operations. Contractor personnel need to know where they can safely go and where heavy equipment such as pump hoists are unusable. To the best of their ability, however, site management should work to remove obstacles to permit the contractor to do its work quickly and effectively, saving on billable hours and shortening the time until wells are back on-line.

Site management should keep regular contact with contractor supervisors on-site to check on progress and to assist in any way possible to keep the work moving along. Rehabilitation work may be frustrating to watch because there are long periods of waiting during chemical contact times, mixing, and surging. If people are sitting around at the contractor's vehicles, this is normal and not necessarily shirking, but it is a good time to make contact. Simple pump repair work is more straightforward and should proceed quickly.

Performance verification. After the work is completed at any well, site management or consultant personnel should verify the results and evaluate them in relation to the scope of work and specification goals for performance improvement.

This is then a suitable time for conference to determine if the performance has met the specification goals. Failing to meet goals, or making things worse, in many situations is nonperformance; but in well rehabilitation, it is often difficult to determine all the causes of well deterioration and foresee the problems that may occur.

Rehabilitation Contractor Considerations

Rehabilitation or well service contractors have a different problem: making these difficult jobs profitable and worthwhile. Mirroring the "homework" tasks of the site manager or consultant, the contractor should be fully prepared to *safely and effectively* take on the well rehabilitation tasks under sometimes difficult environmental site conditions.

Safety: What the Contractor Needs to Have and Know

If they are planning to tackle pump or well service work on potentially hazardous sites, contractors need to put workers and supervisors through OSHA 40-h training, 8-h supervisor training for any person that will supervise another, and 8-h refreshers. This is expensive, but absolutely necessary to walk on many sites. Contractors should be aware that OSHA will require more record-keeping, state worker compensation insurance may go up, and that EPA personnel on-site are deputized to report to OSHA as well, so scrutiny is possible at any time. (*Readers outside the U.S.: translate this into your own regulatory environment.*)

Contractors should have a general site safety and hazard communications plan written and on file, and be prepared to make it available to potential clients and insurers. In addition, a contractor should draw up a site-specific safety plan for environmental site jobs, which should also be available to all site personnel and site management (and review by OSHA inspectors).

Contrary to some recommendations, most site managers prefer that the site-specific plans be relatively brief and clearly written. The general safety and hazard-communication ("hazcom") plans should be more lengthy and comprehensive.

Well rehabilitation even of potable water wells naturally involves a number of physical and chemical risks. Handling contaminated groundwater and the interactions between rehabilitation chemicals and groundwater constituents involve additional inhalation and dermal contact risk.

In the water well industry, people die every year due to suffocation in hydrogen sulfide atmospheres, and others are burned or killed in methane explosions or due to suffocation. Employees must be informed, equipped, and well trained to deal with the site hazards. Regardless of the legal requirements, this is a trust obligation to valuable trained employees and their families.

Drawing up general safety and hazcom plans. This is well covered in good OSHA site-supervisor courses and in a variety of literature. Contractors contemplating work on environmental hazard sites should have at least the following literature on hand for ready reference (most supplied in supervisor training):

1. *Occupational Safety and Health Guidance Manual for Hazardous Waste Site Activities* (DHHS/NIOSH Publication no. 85-115), National Institute for Occupational Health and Safety, Cincinnati, OH (call 1-800-35-NIOSH). This publication, or its successors, if any, along with your supervisor training, virtually guide you through the writing of your safety plans.
2. *NIOSH Pocket Guide to Chemical Hazards*, National Institute for Occupational Health and Safety, Cincinnati, OH. This publication or its equivalent helps in assessing personnel exposure risk and problems with incompatible chemicals such as the strong oxidants used in well rehabilitation.
3. Multiple copies of *Protecting Health and Safety at Hazardous Waste Sites: An Overview* (EPA/625/9-85-006), U.S. EPA. This is a condensation of the Guidance Manual aimed at "nontechnical personnel" (whoever they are) and good for handing out to employees. If the EPA does not have them, institutions that provide 40-h training often do (e.g., The University of Findlay, Ohio, Hazardous Materials Management Program, 419/424-4647).
4. 29 CFR 1910 and associated OSHA regulations with applicable amendments and corrections, as published in the *Federal Register*, probably as supplied by your 40-h training program, and available from any Federal Bookstore. These contain all the applicable requirements and definitions for "easy reference."

5. Material safety data sheets (MSDS) and other documentation (such as NSF certification) for chemicals used in well treatment. These should be up-to-date and on file, and made available to site management along with your site safety plan. It is sometimes a pain to get *useful* MSDS sheets on some "proprietary" chemical mixtures, but you should *insist that suppliers provide them or that you will buy materials from someone else who will.*

How are these health and safety plans written? If a contracting firm is typical, writing is not the first love of its key people, and writing a suitable plan may seem to be a tremendous task ranking in preference well behind stripping and repainting a rig. If this is the case, the firm is well advised to hire a technical writer to perform the task. Such writers are available around the country, many of them with experience with environmental documentation. The Society for Technical Communication (703/522-4114) or industry professional associations such as the National Ground Water Association can provide reference to suitable writers in a contractor's area. There may even be a few writers with both environmental safety and well rehabilitation experience. It pays to find out.

Why not just borrow the consultant's plan? There are some drawbacks:

1. Assuming they will share it, the contractor does not know whether the consultant's plans are fully accurate, suitable, or in compliance with regulations.
2. Many aspects of the consultant's plans are not applicable to the contractor and vice versa.
3. Finally, these plans may have to stand up to scrutiny by unfriendly people, and it is not a good idea to possibly be at the mercy of the engineer trainee (however diligent) who may have written the consultant's plan.

A technical writer working directly for contractor management will do a better job at typically less cost. The writer's work should be reviewed by an attorney competent in environmental health and safety regulations.

Practical Stuff: Access and Response

Problems abound in well and pump rehabilitation and repair on hazardous sites. Some of these jobs best resemble undersea repair by divers. Access is usually a problem for heavy equipment, and pumps themselves may be hard to reach or retrieve thanks to engineering that never considered that eventuality. Wells may be anywhere depending on need, so the contractors may find themselves rehabilitating a well down a bank in a "wetland" 100 ft from the "road."

Going out for parts may be very difficult and waste hours if decontamination and unsuiting are necessary. In these situations, contractors really have to plan ahead or have people on the outside standing by. *It helps if contractors have on hand accurate equipment and configuration descriptions, equipment and part numbers, and well locations.*

Contractors and Consultants: Avoiding Trouble in Working Together

There are a number of considerations for consultants and contractors to assure a harmonious relationship. Areas of friction occur most frequently in contract and specification language, specifically regarding environmental responsibility, costs and fees, and definitions of performance.

Mutual Respect in Rehabilitation Work

Contractors should also avoid bypassing or duplicating the work of consultants for the sites in question. Environmental control (e.g., landfills) and remediation are usually managed by multidisciplinary environmental engineering or management firms, and any trespassing on their "turf" can result in a loss of good contract potentials. A better approach is to establish mutually beneficial relationships with such firms, allowing both to profit reasonably from their respective skills and assets.

As in monitoring and other site contract work, well rehabilitation contractors should have a realistic view of the inflated costs of doing business in this environment that also go along with the higher profit potential in environmental work.

Specifications: Business and Bidding Considerations

Specification pitfalls. What jobs will competent contractors bid on, or avoid with a 10-ft (or 10-m) pole? Well rehabilitation specifications that contains the following elements deserve to be avoided or modified in negotiation.

1. Specifies in detail the exact methods to rehabilitate, but...
2. Fails to provide sufficient information on the wells to be rehabilitated.
3. Specifies what the contractor knows from experience will be inadequate rehabilitation methods.
4. Sets precise performance standards for acceptance of work ("...401% increase in specific capacity...").
5. Sets unreasonable time limits.
6. Specifies unreasonably low or rigid cost caps.
7. Contractor liability for purge material or anything else beyond the scope of work.

From the consultant or supervising engineer's standpoint, this can be a very demanding list of caveats. Well rehabilitation work best resembles site drilling work in that it is (ideally) a service provided by highly skilled technical professionals in an atmosphere (usually) of uncertainty and limited information. It is different from drilling because site supervisors probably have less experience with well rehabilitation than drilling, well construction, and environmental testing, and there are no established standard guides for performance.

Overcoming pitfalls. The experienced well rehabilitation contractor probably has more specific knowledge than the site supervisor's personnel of the technical issues in the rehabilitation task: chemical interactions, volumes, time, surging, equipment, etc. The supervising professional should acknowledge this and take advantage of the experience and knowledge the contractor provides. For example:

1. *Let them take the lead or have significant input into the rehabilitation plan.* Engineers may prefer precise control of such crisis projects and thus prefer providing detailed instructions to contractors. However, the more specifically experienced contractor may have recommendations that boost effectiveness, save time or money, or improve safety. These should be entertained and incorporated as feasible.
2. *Let the contractors have the information they want to plan the job well.* Information may be sensitive or may not be in a form easily transmitted in a specification. Good well rehabilitation contractors or consultants demand all the available information on the wells in question and all potential hazards to be encountered. They will be able to sift through what the site management has.
3. *Give them flexibility in doing the job.* No one wants costs or schedules to balloon out of control. However, it is very difficult to precisely define costs and time with the uncertainties of well rehabilitation. There has to be trust and communication to provide reasons and reassurance.
4. *Avoid precise performance goals.* Such goals are difficult to define. There are numerous factors that may prevent well rehabilitation from achieving desired improvements. On the other hand, contractors must be prepared to give and defend good reasons why performance goals were not achieved.

Specifications and contracts for pump service work can be more narrowly specific in scope and price structure because the engineers set the pump requirements. Hours and other contractor costs are also likely to be better defined, more like electrical and plumbing contracting. For more technical considerations, see Specifications for Rehabilitation.

Purge Water Containment

Well rehabilitation usually generates purge water that may be defined as hazardous wastewater. An issue is, "who should be responsible for control of this water?"

If the site is already involved in containing contaminated soil or water, or hazardous substances, site management has already made arrangements for secure disposal or treatment. However, if the contractor offers fully qualified hauling to secure disposal or on-site treatment (at a price), this is a good choice. The party responsible for the contaminated material (the agent of the property owner) should receive copies of chain-of-custody documents from the hauler.

WELL REHABILITATION: DECISION-MAKING ON METHODS

To Rehab or Not to Rehab — That is the Question

The first issue to resolve: Rehabilitate these wells or not? Several questions can be answered:

How intrinsically valuable is the well? Value for monitoring or remediation wells is subjective and can be defined both in terms of replacement cost and whether the well can be easily replaced as a monitoring or pumping point. What has to be done administratively to replace the well? How much time does that take? The physical act of new construction of such wells is rarely expensive in itself.

What is the local history of well problems? What are the history and structural condition of the well systems? Does the well system have a history of chronic clogging, corrosion, pump failure, sand pumping, etc. Does a well produce noticeably less water than comparable wells in the area or site (was this sudden or gradual)? If low producers are common in the area, do they respond to redevelopment? Another consideration is whether the well meets current technical standards (e.g., D 5092): Is it well constructed, well located, and efficient? If it is not, is it so deteriorated as to be beyond reconstruction?

Is well rehabilitation technically effective? Is it possible to get long-term or permanent well improvement by relocating a well or does the condition reappear in new wells? Can the current well construction materials withstand the chemicals, heat, and physical shock of redevelopment? Is there a chance that redevelopment tools will be hung up, lost, or severely contaminated during redevelopment? The answer to any of these determines the technical viability of rehabilitation (forget cost effectiveness, regulatory constraints or other practice issues for the moment).

If conditions quickly reappear, preventive maintenance has to be practiced to protect the new wells. Rehabilitation alone won't work.

If the well will not stand up to proposed treatment, it may be necessary to: (1) tone down the treatment or (2) abandon a well and start over.

If tools can be lost, damaged, or contaminated so that they are difficult to decontaminate, the first question is: Does the contractor try to perform decontamination? And secondly, will the client buy the tools if they are lost or contaminated beyond reasonable recovery?

How expensive will the well be to replace? In general, relatively shallow monitoring and recovery or plume control wells are not especially complicated or finely designed. However, administrative considerations (especially the costs associated with design time, design approval, etc.) can make each well worthy of saving if possible.

Cost goes up as wells are deeper, drilled in difficult formations, have long, variable-slot screens, or are otherwise more complicated. Larger-diameter wells are of course more expensive per unit depth.

The major problem in monitoring applications may be in nested and other multiple completion wells where one well or interval is a problem, while others are not. To replace such a well may require replacement of an entire nest or disruption of multiple intervals. In this case, an attempt at rehabilitation is preferable.

Will the well be difficult to relocate? Another monitoring well problem is with wells that are critical or hard-to-replace data points, or those that have been monitored for a number of years. If such a well becomes unusable or its data unreliable, rehabilitation is a good option, even if water quality is disrupted briefly, because new well construction is also disruptive, and "paperwork" for new construction becomes a cost and administration issue in many places. However, if the well is substandard or irretrievably deteriorated, then it is preferable to attempt to replace the well in the same location. *In doing so, it must be recognized that the deteriorating conditions may remain to affect the new well.*

Just how bad is the well? Can site management live with the well as it is or not? Historically, experience indicates that well owners and managers will live a remarkably long time with poor wells rather than make the investment in improving them (the curse of the well maintenance consultant and contractor).

For pumping wells, what is the well's working efficiency and what does it cost the well operator to operate at the poor efficiency vs. an improved efficiency? In terms of wire-to-water or well efficiency, using two useful equations from Helweg, Scott, and Scalmanini (1983) provides some information to use in making decisions about changes in well efficiency:

$$E_w = SC_{act}/SC_{max}(100) \qquad (1)$$

where E_w = well efficiency in percent
SC_{act} = current actual specific capacity (SC)
SC_{max} = original or calculated theoretical maximum SC

$$E_n = E_o/E_{max}(100) \qquad (2)$$

where E_n = normalized pump efficiency in percent
E_o = measured pump efficiency in percent
E_{max} = maximum or new pump efficiency

Note: E_{max} should have been tested at the well and not under ideal factory conditions. Efficiencies should be compared at the same total dynamic head (TDH) and flow (e.g., gpm) rate (e.g., 100 ft TDH and 30 gpm), not E_o at 100 ft and 300 gpm and E_{max} at 50 ft and 250 gpm. E_n should not be used alone as an indicator of well efficiency because it is only a pump efficiency value.

Actual calculations of cost effectiveness require a knowledge of the costs of operation at a given efficiency. One cost consideration equation is provided in the following section. It is important to note that such equations do not directly factor diminished screen effectiveness or corrosion wear, but these will be reflected in reduced performance efficiency.

Reduced pump service life or more frequent repair intervals are significant factors in operational costs. If a $2000 pump lasts only 2 years (the warranty period) before replacement instead of 5, the $2000 pump is purchased 2½ times instead of once.

Other matters directly considered are the tangible and intangible costs of poor water quality, and fouling of water treatment and water system components.

The Sutherland et al. (1993) or similar spreadsheet cost analysis approach expands the scope of cost analysis.

The Costs of Well Rehabilitation

Well rehabilitation decision-making processes fall into three categories:

1. *The "well cleaner's dream":* The well in question is irreplaceable, cannot be operated as it is, and new well construction or other options are not alternatives. In this case, cost effectiveness will be no real factor in the decision to rehabilitate.
2. *Looking at rehabilitation vs. doing nothing:* The natural inclination of the well operating management is to do nothing until a crisis develops. It may be up to the seller of well rehabilitation services or the well system operator to convince the site management of the actual benefits of rehabilitating the well.
3. *Rehabilitation vs. new construction:* To drill or to analyze the problem and solve it? Assuming that the well will stand up to rehabilitation and cleaning, and it is otherwise feasible to rehabilitate, the cost of the rehab job determines the decision in most cases. If the cost approaches the cost of new drilling, the owner will be inclined to drill new unless there are other constraints.

The Cost of Doing Nothing

Usually this is expressed in the increased cost of pumping water, either due to increased well losses, expressed in increased drawdown or in reduced wire-to-water efficiency (e.g., Helweg, Scott, and Scalmanini, 1983):

$$C = \frac{Q\,(s + SWL + P)\,(0.746)\,(T)\,(K)}{3956 \times E_o} \tag{3}$$

where C   = total cost of operating over time
       Q   = discharge in gpm

s   = drawdown in ft
SWL = static water level in ft
P   = system pressure in ft of head
E_o = overall efficiency as a decimal
T   = time pumped in hours
K   = cost in dollars per kWh
0.746 = conversion factor, hp to kW
3956 = conversion factor, gpm × ft to hp

Example: $Q = 100$ gpm, $s = 30$ ft, SWL $= 100$ ft, $P = 115$ ft, $E_o = 0.60$, $T = 24$ h, $K = \$0.07$.

$$C = \frac{100\,(30 + 100 + 115)\,(0.746)\,(24)\,(0.07)}{3956 \times 0.60}$$

$$C = \frac{30,705.36}{2373.6} = \$12.94 \text{ per 24-h day to pump that well.}$$

Change s to 50 ft, and C = \$13.94 per day
Change E_o to 0.50, and C = \$15.52 per day

For 24 h/day, 365 days/year pumping, \$15.52/day = \$5664.80 per year, vs. \$4723.10 at \$12.94/day. The nearly \$942.00 difference on this one well could buy a lot of electricity. The cost effect is more dramatic for steeper efficiency losses and larger, deeper wells, and multiple-well systems, and less dramatic for smaller, lower-flow wells and systems.

The next step is to convince management that the cost of the rehabilitation (usually more than the \$942.00 per well) is worth the effort. Here, it is necessary to look at (1) the long term: convincing management that, with regular maintenance, the well cleaning will pay for itself in power savings over a period of years; and (2) other factors beyond mere power costs, such as sandpumping and bacterial fouling of filters and treatment tanks, which affect critical site remediation performance and increase the cost benefit. This more dramatic expression of well problems is more common in environmental remediation extraction wells.

Costs for Serious Rehabilitation Work

Historically, this ranges from 10 to >100% of new construction. It is closer to 10 to 20% of new construction for municipal and irrigation water supply wells and closer to the 100% for environmental site wells, excluding design, planning, and other associated permitting costs.

The 20% figure is a psychological barrier for many in well-owning management (even if told that new construction is not a permanent solution). Above 20%, drilling new wells seems to be the more attractive option if feasible. Sometimes the options are limited. For irreplaceable wells, prices can be 100% of new drilling if the result is the same: an operational well providing the necessary flow.

The cost in any case is higher if the well is more deteriorated or plugged, or harder to clean for whatever reason, vs. a light-weight problem. *The incentive here for management is, then, of course, to make sure that wells do not become so radically deteriorated.* Wells are like any other working machine or structure in this regard.

The potential for abuse can be great in pricing (or slicing of service if there are price caps) because it is difficult to monitor well rehabilitation progress while the rig is over the hole. However, this is a business trust issue, not a technical one, or a reason not to rehabilitate. It simply means that there has to be trust, honor, and accountability in the relationship.

Note: Trust and confidence in well rehabilitation contractors has been a running theme here in (1) site safety, (2) participation in well M&R planning and flexibility in implementation, and now (3) costing and value for the money.

How does a manager gauge the qualifications and trustworthiness of a contractor? Management should check into past experience and referrals from other clients in choosing appropriate contractors. Ask for the bad as well as the good. Call other clients of this contractor and check into problems with performance, billing, timeliness of service (keeping in mind that the person on the other end of the phone may not have a true sense of the time problems), follow-up, and general professionalism.

If negatives come up, discuss these with the potential contractor. Do not just write them off without a chance to explain. Examples are given below.

Why were you 4 weeks late in completing that job? Answers: Our chemical supplier was backordered on the necessary chemical and we could not find an alternative approved by the state DEP. Site conditions were too soft to risk bringing in our equipment.

Why did that job cost 150% of estimates? Answers: Our chemical costs went up since that acid is no longer a process by-product. We had to make special wellhead modifications to permit us to maintain a positive head during chemical introduction, a change condition on that job, the contractor adds.

The project manager at XYZ says you left the job site a mess and were rude and disrespectful to him. Answer: They insisted we start on that site in the Ohio Valley in March and we had to winch in and winch out. We're sorry, but tank trucks and pump rigs are heavy! We're sorry we were rude, but he was always in the way telling us what to do... OK, some jobs just do not go well, do they?

If trust and confidence in current contractors is strained, problems should be resolved by discussion and changes made if problems are not resolved satisfactorily and trust cannot be reestablished.

Contractor Pricing of Rehabilitation Work

The trick for the contractor is to provide a viable service at a cost feasible and salable to site management, while also making a living at it. Some factors include:

1. *Costs:* Capital equipment (spudder, compressor, jetting rig, and tool strings with special development tools), chemicals, fuel and power for engines and boilers, most especially labor and overhead. Insurance, training, travel, and mobilization-demobilization costs are very important expenses. There is no avoiding some significant expense, but costs can be kept down with some basic ingenuity and frugality. Some existing equipment laying idle can be recycled into rehab at little cost: spudders (cable tool rigs), low-pressure compressors, pump hoists, drilling pipe, etc. Chemical handling equipment is low cost. The chemicals themselves may be cheap to expensive vs. unit volume and type (getting more expensive). Packers and high-pressure pumps are relatively expensive, and there is not much to do about that but to take care of them. Training and protective gear to handle chemicals are expensive, but absolutely justified.

2. *Pricing (in general):* The typical costs (e.g., 16 man-days, 1600 gal of chemicals, travel, or equipment time and overhead) can usually be accommodated within a fraction of the expected charge. Contractors may suggest a price structure in which the contractor charges that amount "for trying," receiving a bonus for results. Just to illustrate contrasts, on water well hydrofracturing jobs, for example, the costs are so low that $500 will cover the attempt unless there is a lot of travel involved, and charges of $1500 for good results are not unusual. On the other hand, cleaning a line of biofouled plume control wells may run into the $100,000s.

3. *Pricing (on a particular job):* Given what you have been told about the wells in question: Can the contractor afford to do the job? Is it too far away? Will it require extra effort? The contractor might have to decide: are there times when you need a "loss leader" to demonstrate your techniques or to gain experience?

Choosing Rehabilitation Methods

Before choosing a method, it is necessary to analyze the need. It is easy to "overkill" or "underkill" the job. Review the analysis part of this text (Section II) for methods and applications for analyses to determine what the problems are that have to be treated. The rehabilitative methods themselves are really only variations on water well methods. All are undergoing some development and improvement over time. Several are described in Chapters 9 and 10.

All active rehabilitative treatment strategies (i.e., excluding redesign and material selection) in practice involve some combination of removing encrustants, suppressing microbial growth, and clearing the material from the well.

All such methods should be considered temporary measures. Maintenance follow-up is required to assure continued improved performance. Maintenance follow-up should always be part of the project discussion.

Issues in Rehabilitation Chemical Selection

Rehabilitation chemical selection for environmental applications is more restrictive than for water well applications (at least so far). In water wells, the criteria are that the chemicals (1) are effective for the purpose, (2) have low toxicity at low residual concentrations (chlorine but not formaldehyde can be used for example), and (3) are removable in the development process. Reactivity with the groundwater chemistry is also an issue, but only rarely technically restrictive (getting permission is another matter in some jurisdictions).

In environmental applications, chemical reactivity is the prime concern. Acute or chronic toxicity are less important than in water supply or discharge because the chemical mixture resulting from the mixing of treatment chemicals and groundwater will be contained, and not usually handled by the well cleaning crew. Effectiveness in dissolving and dispersing fouling material has to come in third.

For purge and recovery wells, reactivity is an issue because the groundwater is usually contaminated with a relatively significant concentration of some combination of chemicals. They did not put these wells and treatment system there for a design exercise, did they?

In monitoring wells, the chemical treatment should not leave a long-term residual change in the "native" well environment. Actually, the evolving standard for monitoring wells is *NOT to use chemicals AT ALL.* If this status quo continues to stand, chemicals will not be used in monitoring well rehabilitation.

Reactivity is different from place to place, of course, depending on the contaminants present. Some compounds are explosively or thermally reactive with oxidants such as chlorine or hydrogen peroxide.

If the contaminants are primarily organic compounds that do not react violently with oxidants, the issue is the fate of the altered product water. Using chlorine will result in chlorinated organics (disinfection by-products, DBP) that are more recalcitrant for bioremediation, for example. (DBP are also a problem for public water supply systems because long-term exposure to such compounds is possibly carcinogenic.)

Development water contaminated with chlorinated organics has to be contained and treated. Also, if a relatively pure contaminant such as toluene is being recovered from spent filters, then the presence of chlorine contaminates the product and the filtrate is less valuable or useless. Such chlorinated organics should be removable in carbon filtration or air stripping, or by reductive dehalogenation, however.

Assuming that reactivity and purge water handling are not problems, chemicals should be chosen based on effectiveness. Within the limited list available once chemicals are eliminated for reactivity and toxicity, chemicals used in well rehabilitation are chosen according to the nature of the well problem, which has been determined by analysis, not guessing. There are no standard or stock procedures.

Again: Aside from safety issues, chemicals should primarily be chosen to assist in removal of clogging and encrusting materials, based on a diagnosis of the problem in the well. And, before selecting a chemical regime, *it is prudent to be thoroughly familiar with modern well rehabilitation methods for these jobs.* Some who are assigned the task of specifying chemical selections assume it is (1) simple or (2) can be made from an old spec, text, or with a phone call. This is really not the case, and one should not bet their job on oversimplification.

Specification writing should not be attempted by anyone not familiar with:

1. Actual well problems (determined by analysis).
2. Effectiveness and reactivity of the chemicals.
3. Practical considerations in application of chemicals.

Well rehabilitation methods are in a period of change as new information improves on the knowledge of the effects of treatments on well-deteriorating problems. For example, heat augmentation may be used to boost the effectiveness of chemicals, and special proprietary blends and methods are increasingly marketed and employed by contractors in North America, Europe, and Australia.

Beyond knowledge of the problem and applications, it is important to note that many of these chemicals are quite hazardous to handle if proper safety procedures are not followed. They should only be used by trained personnel familiar with their safe use and equipped with proper personnel protection (respiratory and dermal). For people specifying treatment with chemicals, this is the ultimate issue (assuming human life is valuable to you).

The project management should know the following. (1) Can the chemicals and their mixture with the treated groundwater be safely handled by the personnel available? Do they have the training, experience, and equipment? If not, do not try to make "90-day wonders." Hire experience. (2) Will the proposed treatment meet with regulatory approval? The mentality in well rehabilitation circles in the past was to "sock it" with strong, highly reactive compounds, primarily "hard" mineral acids and (secondarily) highly concentrated hypochlorites. There was little regulatory oversight. In the past 5 years, regulatory agencies have begun to scrutinize what is being used in well rehabilitation, both for potable water supply and environmental wells. There has now developed a tendency of extreme conservatism and reluctance to approve any chemical use.

Keeping these issues in mind, Chapter 9 is a review of physical and chemical treatment applications in well rehabilitation and maintenance. Many of these are discussed in greater detail by Borch, Smith, and Noble (1993).

Reconstruction

The structure of the well sometimes has to be renewed or modified to solve a problem, usually a "mud seam" or other source of fines, or to repair collapsed or corroded sections.

Usually, for any of these problems, a suitable liner is installed and grouted into place where there is sufficient diameter. Installing liners can be complicated and not always effective, plus being impractical in smaller wells typical of environmental applications. It does not replace good well design and material choice from the beginning, but is often a choice that can be successfully used. One limitation is that well diameter is lost during the lining process, often more than 5 cm. This can restrict pumps or provide complications in future pump service.

Another reconstruction technique to limit or eliminate sanding is the installation of suction flow control devices (SFCD), as previously described briefly. These have been used with success in Europe, North Africa, the U.S., and Mexico to control sandpumping and encrustation problems in wells (see Chapter 7). Further reading on SFCD function and application can be found in Pelzer and Smith (1990), Nuzman and Jackson (1990), Ehrhardt and Pelzer (1992), and Borch, Smith, and Noble (1993).

Specifications for Rehabilitation

Specification writing for well rehabilitation in general is frankly in a disreputable state in the U.S. Locally, there are surely pockets of excellence, but specifications coming across the desk of the author usually have some of the following deficiencies:

1. *No provision for analysis of well performance or, especially, the qualitative nature of the fouling problem.* The treatment method has been preordained by the engineer writer. The bidder therefore has no idea whether the problem has been properly diagnosed, the extent of the problem, or the well history.
2. *There is insufficient information on the pump and well structure for the bidder to make intelligent judgments.* The bidder has to have useful information to make a decision about the vulnerability of the well, chemical types and volumes recommended, or what has to be done to fix or refurbish the pump. These are all factors in chosen approach and cost analysis.
3. *The specification writer has provided an exact procedure and chemical volume specification that the bidder is to follow in exact detail.* The bidder has no idea, of course, if the specified well rehabilitation procedure is a good match without sufficient analysis information. At best, such procedures are usually dated or insufficient.
4. *There are no performance standards or provision for follow-up.* The process specified is assumed to do the job, but there is no budgeted time to find out, and no common ground for agreement between contractor and client to determine effective performance.
5. *There is too much irrelevant detail.* Government RFPs sent out for general bid of course typically devote 100 pages to equality in employment and maintaining a drug-free workplace and 3 pages to technical details. The 100 pages part cannot be helped, of course, due to regulations, but these RFPs could often be more user-friendly and informative about the technical issues. The same is true of specifications written for the private sector.

What Well Rehabilitation Specifications Should Have

Specifications for well rehabilitation should have several features, including:

1. Cause of the well problem has been or will be properly analyzed (information available to the bidder).
2. Treatment specified can be adapted to address the specific problem if necessary.
3. To be qualified, the treatment contractor has to be knowledgeable and adaptable (and should be able to demonstrate these qualities).
4. Well will be tested before and after for performance.
5. Well will then be maintained properly to prevent or mitigate recurrence of the problem.

Item 1, Problem analysis: Specifications for the whole project (if not the well cleaning contract itself) should require microbial, physicochemical, and performance analysis (as recommended in Section II), and TV inspection (where possible), before and after treatment. Such tests should be diagnostic and practical in nature.

Item 2, The treatment specification should be goal- and not process-oriented: The goal is increased performance, reduced fouling, etc. The exact treatment can be left to the knowledgeable contractor. The contractor should be required to submit a description of the proposed procedure for assessment by a knowledgeable project manager or consultant.

Item 3, Limits to contractor freedom: The major exceptions to the freedom of item (2) are (a) the presence of contaminants that react poorly with major well cleaning compounds and (b) the site and well history are so well known that the treatment can be narrowly defined based on past experience. Chlorine may be banned, for example, and a replacement recommended (based on experience and not on a chart).

Other exceptions to (2) are (a) an example treatment protocol to permit "level-field" bidding by all interested or (b) a protocol highly specific to a favored contractor who has provided excellent service with a specific approach. By "service," what is meant is service at the wellhead, not the one with the best advertising visibility or most persuasive sales representatives.

Item 4, Performance testing: There should be some suitable testing procedure to determine the extent of difference between the condition before and after rehabilitation to rationally assess the effectiveness of the treatment. Generally, Item 1 above defines the "before" and similar methods are used to define "after."

Item 5, Post-treatment: The specification should have provision for providing follow-up maintenance. The project may make this part of the well cleaning package, or separate.

Specification checklist: The specification should ask for:

1. Contractor's recommended approach and fees based accordingly, spelled out intelligently.

2. Contractor company and personnel qualifications, experience, training, and references.
3. Assurance of contractor training, licensing, etc.

Selecting Well Rehabilitation Bids

Bid proposals should be accepted based on the qualifications and demonstrated competence and knowledge of the bidder, as well as price. Bids accepted should be "best." Of course, price is important, but only if there is value for the price. Low bids that provide substandard service may cause greater life cycle cost in terms of: (1) reduced pump life, (2) well damage, (3) too-rapid return of the offending condition, and (4) higher pumping costs due to lower than necessary efficiency.

Experience and reputation in this business are important. Bid reviewers should themselves be knowledgeable about the issues, and judge experience and reputation based on actual performance in the recent past. Separate this from (1) decades of company existence (the legendary old experts extolled may now be dead) and (2) marketing atmosphere and sophistication.

9   Rehabilitation Methods: Technical Descriptions

Methods differ depending on the circumstances, but this is a description of methods for general information. For the most part, these are suitable for maintenance as well as rehabilitative treatments within limits.

PHYSICAL AGITATION

Physical agitation is recommended for most rehabilitation and maintenance treatments for the same reason that when you wash a bottle, you scrub it out — agitation makes the washing more effective. In many cases, the chemicals are relatively ineffective against entrenched deposits in the cold groundwater. Agitation methods can take several forms.

"Conventional" Redevelopment

Redevelopment is usually required for most well cleaning — surging with surge block or air, or jetting for redevelopment. Brushing is an important and useful variation for removing encrustation from shutter screen and open rock boreholes. See the discussion on well development methods in Chapter 4 (virtually the same procedures) and also other relevant references (e.g., Borch, Smith, and Noble, 1993; Roscoe Moss Co., 1992; Driscoll, 1986; Standard Guide D 5521).

One important difference is in handling development tools if they are being used with hot water, cryogenic carbon dioxide, or corrosive/caustic chemicals. Chemical-laden splash and discharge have to be tightly controlled. The development tools themselves have to stand up to the chemical or thermal environment. Personnel have to be knowledgeable and respectful about handling chemicals and hot (or supercold) water and tools, and equipped to do so safely.

"Other" Physical Redevelopment Methods

There are a variety of additional physical development methods used for water supply, oil, recharge, and other wells, including shooting and hydrofracturing.

(Refer to Borch, Smith, and Noble, 1993; hydrofracturing is further described by Smith, 1989.) Many of these methods, including hydrofracturing, are probably too vigorous and not sufficiently controlled for environmental well redevelopment, especially where contaminated water is involved. However, existing hydrofracturing equipment can be used for supplying pressure to jetting tools.

There are two such physical development methods that have the potential for usefulness in rock and more robust screened well situations.

Cold CO_2 "Fracking"

The Aqua-Freed™ procedure (Aqua Freed, Newburgh, NY) was developed as a way to provide the redevelopment effects of cryogenic CO_2 in a controlled manner. This process has four steps, as follows:

1. Injection of cryogenic CO_2 under pressure.
2. Allowing time for penetration into the formation.
3. Repeating as necessary.
4. When desired, venting and depressurization.

This process is described as acting on the formation and encrustants in the wells through freezing and thawing and thus dislodging deposits, and also through the formation of carbonic acid, acting under pressure. The carbonic solution is relatively high in concentration and acts as an acidizer. The thermal shock on bacteria and their ECP probably has some benefit in dislodging biofouling.

The Aqua-Freed process may have promise in the environmental field for several reasons.

1. The injectant is chemically reduced, not reactive with organic molecules, and does not stimulate respiration.
2. It does not work under high pressure, so that fracture opening is minimized.
3. The material, compressed CO_2, is relatively safe to handle (it is however supercold), and no other chemicals are considered necessary for the method to work.

One current drawback is the limited distribution of this treatment process to a few well service contractors (access is tightly controlled) so there is little existing price competition. Another drawback is using this process in wells of limited physical integrity. Instances of jacking casings out of the ground or extruding bentonite annular grout have been reported by the contractors. Resolving such problems should be a matter of adjusting procedures to meet the limitations of the wells. However, before proceeding, potential clients should have a candid discussion of the possibilities with contractor representatives. Independent evaluation of the method's effectiveness against various well problems is also not yet available.

Sonic/Vibratory Disruption

Some sonic or vibratory methods have been developed to provide controlled shock treatments for wells. One is the Sonar-Jet™ treatment (Water Well Redevelopers, Inc., Anaheim, CA), which produces high-velocity shock waves in the well. This method is particularly useful for removing hard deposits on and around louvered gravel pack screens and borehole walls. Sonar Jetting is more safe and more controlled than blasting. Currently, one problem is that this technology may not be available everywhere, although it is more widely available than before.

Sonic methods to date have little effect on soft deposits. They are, however, useful in removing hard deposits to permit access to biofouling deposits. A revised treatment protocol (see following section on chemical applications) involving a sonic step is:

1. Pretreatment with shock chlorination or disinfection (see following discussion).
2. Treatment with vibratory method.
3. Pumping to waste.
4. Shock disruptive chemical step/agitation/pump off.
5. Acidization/agitation/pump off.
6. Shock disinfection/agitation/pump off.
7. Test for effectiveness.
8. Repeat as necessary.

THE PHARMACOPOEIA: CHEMICAL USE IN REHABILITATION

The chemicals mentioned here represent many of the available options. They will be summarized in terms of both effectiveness against various well problems and their environmental reactivity and other such practical problems.

They may be classified in a number of ways, but may be broadly considered as (1) acids, used in dissolving and disruption of deposits, (2) antibacterial agents (used for disinfection and control of biofouling — there is some overlap with acidizers), and (3) sequestering agents, used to aid acid and disinfectant penetration and to remove purged deposits (some of these are organic acids).

Within these classifications, particular agents can be further classified as: (a) *"Generics"*: these are the chemical mixtures as available industrially without special blends and suitable for well cleaning; and (b) *"Alphabet brews"*: numerous proprietary chemical products are marketed for well cleaning. They are mostly enhanced mixtures of common acid compounds; but whatever their purpose or class, these should be evaluated based on the information provided by the supplier (although this is often sketchy), and recommendations of a well maintenance consultant and other users.

Where possible, proprietary compounds will be discussed here based on their method of action (acids, etc.).

Note: In monitoring wells or elsewhere in which chemical introduction is excluded by regulatory or project managers, the following information is useless. While chemicals may be completely removed in such cases and the subsurface environment returned to "normal," management often considers the risk too great to gamble on. *If this is the case, the monitoring wellfield manager had better have a good maintenance program or performance decline will occur without any effective means of restoring performance.*

Acidizing

Acids are used to remove blockage in carbonate rock aquifers, dissolve iron/manganese oxides and carbonate encrustation, and have some antibacterial effect by providing an ionic shock to aquifer microflora typically adapted to circumneutral pH. Not all acids are equally desirable for the purpose. In wells and water systems, the acids historically and most commonly used in rehabilitation are muriatic (industrial-grade hydrochloric acid, HCl), sulfamic (H_3NO_3S), and hydroxyacetic ($C_2H_4O_3$) acids.

Acid Properties

Muriatic acid is one of the most powerful acids used for removing mineral scale and comes in liquid form. It can be purchased with an inhibitor that minimizes the acid's corrosive effect on metal well screens, casing, and pump components.

Although it is often an effective well cleaner, it is not especially effective against iron biofouling (the premier problem in monitoring and extraction wells) and is hazardous to handle. Respirator and full-body splash protection is required and washdown treatment facilities should be immediately available. Once placed in the well, toxic fumes are expelled from the borehole within moments. Inhalation of these fumes can cause death, and contact of the liquid with human tissue can result in serious injury. On an already difficult environmental job, with personal protection, etc., the hassle of dealing with muriatic acid is an unnecessary additional worry.

Muriatic acid presents purge water handling problems as well. There is a tendency to overdose based on common tables, so that much acid is wasted, and these doses depress pH to <1. If it is not extensively buffered or neutralized by substances in the well, the purge water is highly corrosive and difficult to dispose of properly and safely.

Muriatic acid, as generally available, also has a reputation of containing arsenic and metal contaminants. The resultant purge water brew may be unacceptable to wastewater treatment or disposal.

Sulfamic acid is a dry granular material that produces a strong acid when mixed with water. Sulfamic acid should not be confused with sulfuric acid (a yellow liquid), which should never be used in well cleaning, due to the formation of insoluble products.

Sulfamic acid is marginally more expensive than muriatic acid (per volume acidified H_2O solution) and is slightly less aggressive. Sulfamic acid, as a solid, is relatively much more safe to handle and easily transported. The dry material does not give off fumes and will not irritate dry skin on brief contact. If a spill occurs, it can be easily and safely cleaned up. For all these reasons, sulfamic acid is often preferred as an acidizing agent.

Under certain circumstances, dry sulfamic acid can be effectively placed into the water column in carefully measured doses. Alternatively, to produce a measured dosage during jetting into a screen interval, the mixture may be accomplished at the surface.

During in-well treatment or surface mixing, the slowly dissolving acid releases dangerous fumes at a relatively slow rate, so proper ventilation should be always provided.

Less corrosion of pumps, screens, and casing will occur when an inhibitor is added to the acid (some brands have a premixed inhibitor). Little corrosion results when stainless steel well screens are treated repeatedly with inhibited sulfamic acid.

Sulfamic acid has limitations in dealing with certain sulfate compounds. It does not dissolve them well alone and can be very slow with iron and manganese scale. It should be replaced in these situations with other acids more appropriate for the mineral to be dissolved.

As a commercial-grade material, sulfamic acid also has impurities in small concentrations. There is a reported tendency to produce ammonia upon mixing with water under certain circumstances, so its use was blocked in at least one state on an environmental project. All sulfamic acid stock currently originates in east Asia, and there are only feeble impurity standards in force there. As with muriatic acid, pH is greatly depressed at typically recommended mix concentrations. Overdosage is a common problem and purge water may have to be neutralized prior to treatment or disposal.

Hydroxyacetic acid, also known as glycolic acid, is a liquid organic acid and available commercially in 70 to 95% concentrations. Although it is not as well known or commonly used as either muriatic or sulfamic acid as a well treatment, its use has achieved excellent results in well treatment for iron biofouling and has good features for cleaning environmental wells. It is relatively safe to use because it is relatively noncorrosive, nonoxidative (it is an acid, a net H^+ generator), produces few toxic fumes, and is spent quickly. Still, it is a powerful liquid acid, so caution and splash protection are necessary. It also must be labeled to be transported like other liquid acids. Hydroxyacetic acid is a major active ingredient in certain commercial antibacterial acid mixtures. At least one of these (LBA, CETCO Inc.) is NSF listed under Standard 60.

Due to its bactericidal and metal chelating properties, hydroxyacetic acid is often very effective in treating wells with iron biofouling problems. Since hydroxyacetic acid is weaker than hydrochloric and sulfamic acids, its effective pH in solution is higher (pH 2–3), so a longer contact time, or admixture with sulfamic to reduce pH, is required to achieve the same amount of scale removal.

On the other hand, purge water problems are less severe because the pH of spent purge water is higher even if overdosing occurs.

Other organic acids have found use as chelating agents in dispersing encrustants in well cleaning. One acid, oxalic acid, is also effective as a primary acidizer in low-Ca water. It has also been used in wells to attack biofouling, a situation in which the Ca^{2+}-ion concentration problem is less of an issue. Impurities in oxalic acid (all originating in the People's Republic of China) may also be an issue that should be explored.

Using Acidizing in Well Treatment

Acid dosages need to be introduced into the well and forced out into the near-well environment to be effective. The acid mixture may be agitated by surging with a surge block or by cycling the pump irregularly ("rawhiding"). Displacement, pressure injection, or jetting may also be used.

Displacement: After an acid solution is placed in the well or the pellets dissolve, a volume of water equal to that standing in the well screen is poured into the well to force the acid solution through the screen-slot openings into the formation if possible. The displacement water should be introduced slowly to make sure the strong acid mixture does not flow out over the top.

Pressure injection: Injection through a packer is an alternative for relying on head for force. It provides a controllable force and may force the chemical further into the formation. It should be used with caution on bentonite-grouted wells, those that are structurally suspect, and smaller wells in general. Acid migration should also be as tightly controlled as possible.

Jetting: Jetting is another means of controlled injection in which the acid can be directed at specific screen sections. The crew doing the jetting should be skilled and very careful to make sure that there is no splash.

In both pressure injection and jetting application, crews should very carefully connect and inspect hoses and all fittings, and to cable all hoses to prevent disconnection, hose breaks, and resulting hazardous acid spray.

Inhibitors: These are often recommended to limit attack by strong acids on metal in the well. Several types are available, including Rochell's salts and gelatin. Most of the inorganic inhibitors are toxic and must be completely removed after treatment. Gelatin has the unpleasant side effect of sticking to formation materials and providing food for bacterial growth. It also must be completely removed or avoided if this cannot be assured.

Post-treatment for acids: After mechanical agitation, the solution is left in the well to react with the encrustants until a pH of 6.0 to 7.0 has been reached, if possible, then agitated again and pumped to waste. The time for the reaction to occur varies from a few hours to more than 15 h, depending on the type of acids used and the solution concentration, temperature, and the type and amount of encrustants.

The chemical and materials removed during treatment should be pumped off until the product water is clear and close to former quality. Treatment purge water should be disposed of in an environmentally safe manner. Acidic purge water can be neutralized in a lime-filter basin or tank and pumped to wastewater treatment or containment as permitted or required.

If conditions indicate the need, the chemical treatment may be repeated.

Sequestering

Sequestrants act to lower surface tension, solubilize, and wet affected compounds such as metal oxides. In wells, this may be accomplished by application of long-chain polyphosphates, and polyacrylamides and polyelectrolyte mixtures (e.g., Arccsperse™, ARCC, Inc.) that help to remove clays and loosened slime and encrustation. Hydroxyacetic acid, citric acid, and some proprietary acid formulations also have related chelating properties.

Polyphosphates (including brand-name formulations used in water treatment) are losing favor quickly for dispersing biofilms in wells because the chemicals themselves are difficult to remove, remaining behind in the formation to enhance growth, usually at the edge of development influence.

Phosphate is essential to all cells (microbial or the reader's) because it is used in making ATP, the "energy currency" of cells. P is usually a *limiting nutrient* in groundwater for microbial growth because it is typically naturally available only in very low levels or absent (having been siphoned off long ago by surficial soil communities). Adding P-containing compounds can trigger a microbial "bloom." It has to be apparent that adding phosphates improves the chances of aggravating biofouling.

The typical reaction of well performance to introducing phosphates is a rapid removal of clogging material and improvement of performance. This increases confidence and casts doubts on nay-saying. Loss of performance then can return quickly as the now P-fortified residual microflora left after well-cleaning can now rebound and reform their biofilms. Virtually solid, low-conductivity cylinders can be formed at the edge of redevelopment force influence in the aquifer around polyphosphate-treated wells.

These warnings are not aimed at the proper use of polyphosphates in water supply system treatment. In a controlled water system setting (usually chlorinated), long-chain pyrophosphates keep Fe and Mn ions in suspension. The problem in wells is that they are *uncontrolled* and microflora-rich environments. Polyphosphates may also have a place in keeping nonchlorinated remediation treatment system pipelines from closing up with Fe precipitation as long as the lines are monitored for growth and appropriate measures can be taken to control secondary biofouling.

Polyelectrolytes provide the desired effects of dispersing clogging deposits and clay/silt build-up without being P sources. These compounds are *recalcitrant* and are not readily attacked by microorganisms. They can be used in biofouled wells as part of a rehabilitative treatment program, but should be developed out and not left in place.

Post-treatment: All polyphosphates (if they are used regardless) and polyelectrolytes should be removed from the well and confirmed by chemical analysis (total P for polyphosphates and specific indicators for polyelectrolytes). After they are removed, the well should be surged and pumped several bore volumes again to remove any traces of product.

Purge water should be disposed of properly in wastewater treatment, containment, or to controlled surface spreading on soil. Phosphate-loaded water discharged to surface waters can cause algal blooms and oxygen depletion, resulting in suffocation of aquatic animals, and unannounced discharge into wastewater treatment is likely to disrupt the balance in the sludge.

All treatments should be concluded with follow-up testing for residual chemicals, contaminant and biofouling bacteria, and changes in chemistry. If results show inadequate treatment results, treatments should be repeated.

Antibacterial Agents

Treating biofouling and chlorination used to be synonymous; but in recent years, alternatives to chlorination have become more prominent. The following are options in use.

Chlorination

Generally, this is "shock" chlorination with high concentrations as a well treatment, which serves to rapidly kill bacteria and disperse slimes somewhat, breaking down organic polymers. It should only be used where chlorination of water constituents is not a safety or disposal problem. Otherwise, alternative oxidants (hydrogen peroxide) or nonoxidizing antibacterial chemicals (e.g., hydroxyacetic acid) should be used.

Concentrations: As high as 500 to 2000 mg/l of introduced chlorine are usually desirable for treating wells severely plugged by biofouling bacteria. The most easily handled forms of chlorine are calcium hypochlorite (solid, granular) or sodium hypochlorite (liquid solution). Chlorine gas is very effective when used by those trained to handle it safely. Cl_2 is an acid rather than a base and is more effective in shocking biofilms. Again, as with all strong well chemical treatments, this should only be attempted by experts.

Where feasible, shock chlorination should precede any acidizing treatment used for bacterially encrusted wells. Shock chlorination alone (with surging) may be sufficient temporarily for small wells and wells with soft biofouling problems.

Chlorine application: After the chlorine solution has been introduced into the well, it should be forced through the screen-slot openings or water-bearing fractures into the water-bearing formation by adding water to the well (displacement), pressurizing, or jetting as described above for acidizing. Then, as with acid treatment, mechanical agitation should be used to enhance the treatment. As the chlorine disintegrates the organic slime, the mechanical agitation helps dislodge it and move it from the formation into the well where it can be removed by pumping.

Agitation can be achieved by jetting chlorinated water into the formation or by surging in the casing above the screen. Another method is capping the well and alternately injecting and releasing compressed air. Brushing is very effective in helping to dislodge material from the casing, screen, and borehole interior. Chlorine should also be circulated throughout the effected pumping and treatment system (where it does no harm to treatment media).

Longer time intervals between the necessity of treating for iron biofouling in water wells have been achieved by using a three-step treatment consisting of initial shock chlorination, followed by acidizing, and then a final chlorine shock treatment of the entire well structure and connected distribution system if applicable. The added cost of applying three separate treatments is almost always offset by the improved results.

Time always favors chlorine effectiveness; thus, the longer the contact time available, the better. Chlorine solutions should be allowed to work for at least 24 h before being pumped off. If the chlorine residual (free chlorine left after combining with material in the water) is reduced below 50 mg/l, more chlorine should be added.

Post-treatment: After standing, chlorine water should be surged again and pumped off. Purge water with any chlorine residual should be pumped to an open retention pond or tank and the chlorine allowed to dissipate. Alternatively, remove by carbon filtration, which is preferable if the constituents in the purge fluid have to be fully contained. Carbon filtration can reduce the volume of fluid that has to be contained. *Chlorine in even low mg/l concentration can disrupt wastewater treatment and kill aquatic life.* Once dechlorinated, the water should then be discharged to wastewater treatment or containment for secure disposal. If conditions indicate, repeat the procedure.

Although wells can be treated for biological encrustation or biofouling, the bacteria are difficult to eliminate and most problems recur. Prevention and secondarily preventive treatment is the rule for biofouling problems.

Alternatives to Chlorine as Oxidants for Biofouling

With the limitations of chlorination in contaminated environmental wells, several other chemical treatments are possible. These are also primarily oxidants, as is chlorine and its related halogen bromine. Each has benefits and limitations.

Ozone: Ozone (O_3) is formed by exposure of oxygen (O_2) to strong electrical charges. It has to be generated at the point of application due to its instability, which precludes storage under pressure. Ozone is, of course, a powerful oxidant (more so than chlorine, bromine, and oxygen) and biocide.

As a powerful oxidant, ozone aggressively reacts with organic compounds, reducing its availability for biocidal action, and making it dangerous with compounds that are unstable in the presence of oxidants. It does not form halogenated organics, however — a real advantage. Its best application is probably as a preoxidation step in enclosed water system treatment and in combination with other treatments to tackle suspended (planktonic) microflora. Ozone and hydrogen peroxide (discussed next) are used together in commercially available processes (e.g., PEROXONE process; Tobiason et al., 1992).

Hydrogen peroxide: Like ozone, aqueous hydrogen peroxide is a powerful disinfectant and oxidant. It has been used with some effectiveness in removing well biofouling in both water supply and environmental wells. A commercial peroxide mixture is produced by Carela in Germany, but is not currently commercially available in the U.S. A variation is in use in Australia. There are a variety of sources of "generic" 50% peroxide mixtures available commercially in the U.S. and Canada, as well as elsewhere in the world.

Peroxide solutions are volatile liquids that should be handled with the respect given to corrosive chlorine and acid solutions because they can be dangerous to skin and mucus membranes.

As a powerful oxidant, peroxide consequently reacts violently with some compounds. (Check the safety references.) An alternative to handling aqueous H_2O_2 is to form H_2O_2 *in situ*. Sodium perboric (DuPont) can be used for this purpose, but the use of this approach will likewise depend on whether this compound is safe to introduce into the system in question.

It should also be noted that H_2O_2 is aggressively attacked by hydrolase enzymes frequently freely present when bacteria are present. It breaks down to form H_2O and O_2, and the resultant oxygenation can enhance microbial growth away from the well and the lethal oxidant zone. H_2O_2 is, after all, used as a means of providing oxygen in this way for *in situ* bioremediation.

Consequently, such peroxide injection wells for remediation should be relatively trouble-free. However, in the aquifer or wells away from them, biological activity may be intense.

Use of Heat

Studies in Canada in the 1970s have led to a revival of the use of hot water to augment chemical treatments to kill and disperse iron bacteria (a process common, at least in the upper Midwest in the 1920s and 1930s, e.g., used by Layne Northern). Heat and chemicals combined are the bases of two patented well cleaning systems (see following).

In some cases, water heated to 54°C and recirculated over several days was sufficient without chemicals to disperse clays and biofouling at least in the short term (see discussion and Figure 27, Chapter 7). Several contractors in North America routinely employ heated, blended chemical treatments. Heat is increasingly favored as a biofouling removal method where chemicals cannot be used for environmental reasons.

Heat is cumulative, as mentioned before, for preventive treatments, however, and can actually enhance growth away from the thermal shock zone. It is also very inefficient in terms of fuel or power to generate thermal energy. The best approach to using heat is in a process such as the blended chemical heat treatment method described below with a prudent selection of chemicals. If heat is used alone, however, it is important to remember its limitations.

BLENDED METHOD TREATMENTS

One trouble in considering chemical treatment types individually is that individual chemicals seldom do the job alone. Agitation is necessary for chemical treatments to have maximal effect and is the most common augmentation method. Chemicals can be otherwise augmented by mixtures and temperature increase. For example, surfactants improve the contact between disinfectants and bacteria in biofilms, acids provide ionic shock, and such mixtures can be heated to increase molecular activity. Contact time additionally improves effectiveness of biocidal action.

The patented BCHT process (U.S. Patent 4,765,410, ARCC Inc., Daytona Beach, FL) is one good example of the blended method approaches. The BCHT process involves three phases of application to shock, disrupt, and disperse biofouling (Alford, Mansuy, and Cullimore 1989; Cullimore, 1993). It also is unique among available methods in that its effectiveness and results have been studied by unbiased observers (U.S. Army Engineers) on a large scale (e.g., Leach et al., 1991).

The following is one typical scenario. The exact blend of chemicals for a particular wellfield situation is determined based on an analysis of the needs and chemical reactivity among the range of chemicals useful for cleaning the materials, the microorganisms and encrustants present, and groundwater constituents.

In the shock phase, a chlorine solution amended with nonphosphate polyelectrolyte surfactant is jetted into the production zone. The result is (1) a reduction of chlorine demand in the disruption phase, (2) softening of biofouling and encrustants, and (3) increasing microbial kill.

The disruption phase involves more customization (based on analysis of the well conditions), but revolves around jetting in acid blends or chlorine or other biocides blended with stabilized acid, heated to between 60 and 95°C (prior to injection) and allowing a contact time as long as possible. The pH shift is down to as low as pH 1. Heating increases metabolic rates at the fringe of the heat influence zone, increasing assimilation of toxic disinfectants. There is further dissipation of fouling material in the well.

The dispersion phase involves the physical removal of the disrupted fouling material from the affected well surfaces. Standard surging methods are employed.

This method employs all the recommendations of rehabilitative treatment:

1. Analysis of the nature of the problems.
2. Physical agitation in combination with chemicals.
3. Heat augmentation of chemicals.
4. Appropriate mixtures of chemicals customized (based on analyses) for the situation; different acids may be used, and chlorine may be omitted.
5. Staged treatment to produce various effects.

The treatment is followed by analyses of results and treatment is repeated and modified as necessary.

BCHT can have wide application. It has been employed on everything from municipal water wells to redwood-stave pressure-relief wells to pumping remediation wells at RCRA sites. A limiting feature is that application of the process requires very specific knowledge of chemicals, their application, and their effects on fouling, wells, and groundwater quality, and access to the process is limited at the present time. As with CO_2 application, this is not all bad because it means that such treatments must be applied in a tightly controlled fashion by knowledgeable well rehabilitation crews.

APPLICATION OF REHABILITATION METHODS SUMMARY

1. *Preliminaries:* (a) A specific capacity test of the well should be performed before rehabilitation is attempted to benchmark well performance. (b) Bacterial, chemical analyses, and physical well inspection using TV or other methods should be used to determine the cause of the problem.

2. *Redevelopment:* Sometimes, all that is required is to finish the job that the original driller did not complete in the first place. Surging or jetting to remove remaining drilling damage or mud wall, or to finish mixing sediment particles sometimes provides dramatic improvement alone; however, more indepth rehabilitation is usually required. Consult ASTM standard guidance procedures (e.g., D 5521-93), modified for the local situation.

3. *A general well rehabilitation procedure:* While no single procedure is suitable for all problems, a procedure such as this (selected in conjunction with analyses of performance deteriorating problems) will provide some improvement for biofouling and encrustation. *This sample procedure is provided as an example for illustration only.*

a. Test pump well and record performance, physicochemical, and microbial data. Remove well pump and set aside, refurbish, clean or replace.

b. Shock, disrupt and disperse in some fashion using a chemical providing osmotic shock (disinfectant and/or acids) with mechanical surging, e.g., for 4 h, followed by an 18 to 24-h soaking period (may be preceded by nonphosphate dispersant wash with pump running).

c. Jetting or surging with chelating acid solution or further disinfectant with acid for 4 h with pumping. Brushing is often preferable for soft iron build-up on borehole walls.

d. Pump off until well is free of chemicals (use pH and chemical testing methods onsite).

e. Final shock disinfection, methods depending on the volume of solids removed.

f. Inspect for corrosion and wear of well and pump parts, replace defective parts, wash down pump, pipe, wire, etc., with chlorine solution (flush with treated water) or steam decontaminate and reset.

g. Test pump well to determine improvement in specific capacity, physicochemical, and microbial quality.

h. Repeat as necessary.

i. Implement maintenance program (Chapters 6 and 7).

This is one example of a procedure and is not specifically recommended for all purposes.

POST-TREATMENT AFTER WELL REHABILITATION

The natural reaction after a well rehabilitation episode is a feeling of "mission accomplished." However, at least in the case of a chronically biofouled well that has been rehabilitated, the *best comparison is to a human patient who has had extensive open-heart surgery.* The major problems (blockage, encrustation, poor circulation) may be solved for now, but a strict treatment program is now necessary to prevent reoccurrence.

Some Follow-Up "Truisms"

1. *There is no completely effective well rehabilitation treatment known for biofouling and encrustation.* In time, all biofouling and encrustation returns. The purpose of rehabilitation is to restore or improve performance to "buy time" before the problem comes back. On the other hand, redevelopment and reconstruction to control sand or silt pumping often are virtually 100% effective for indefinite periods (for those problems alone, of course).

2. *Maintenance monitoring and treatment as previously described is the only strategy that may prevent a need for further rehabilitation.* It can certainly lengthen the time until full-scale rehabilitation is necessary. Maintenance monitoring and treatment for use after rehabilitation is the same as that described in Chapters 6 and 7. The only difference may be that the testing method and treatments may be more narrowly focused on the problem attacked in rehabilitation.

With biofouling, it is critically important to focus on changing water quality as the indicator of deterioration rather than on loss of hydraulic performance. Remember that changes in water quality happen first.

When testing indicates that biofouling growth is again becoming troublesome, shock disinfection or a staged anti-biofouling maintenance treatment should be employed at the first opportunity. When the treatment is completed, more analyses should be made to test effectiveness and tests repeated as necessary.

Chemical encrustation can be limited using preventive treatments at intervals determined on the basis of testing and experience.

3. *After sufficient experience, a regular treatment interval without much testing is possible, and treatments can become preventive instead of reactive.* A good idea during this maintenance phase is to take steps to reduce the impact of well biofouling, encrustation, or corrosion on the operation of the well and its attached water system. Possible steps:

a. When pump replacement becomes necessary, choose designs that make remediation easier (monitoring or extraction wells).
b. Always use noncorrodible components in remediation pumping wells, such as plastic/stainless or stainless pumps and plastic pump discharge

pipe where feasible. Flexible riser hose such as Wellmaster™ (Angus North America, Angier, NC) permits easy removal of pumps for cleaning. Wellmaster is popular for mine dewatering wells because of this advantage.

c. Check for and eliminate stray electrical currents, as in metallic pipe networks, as much as possible. Cathodic protection may be called for in some industrial settings in the remediation system pipeline network if it is metal. Use plastic pipes as much as possible.

d. Consider the installation of SFCD in wells as much as possible to control sand/silt pumping and to reduce biofouling impact. This buys time until rehabilitation becomes necessary.

e. Install backwashable filters between the well and the distribution system or before the critical part of the treatment system (such as an aeration tower) to limit or prevent the migration of biofouling microbes into hard-to-clean sections.

f. Make the distribution system easy to inspect and clean, for example, by running poly-pigs through the lines.

10   Learning and Going Forward

Experience is the best teacher in many cases. This is an assemblage of relevant case histories (environmental, geotechnical, and water supply) and some final words. Cross-compare with the other chapters.

CASE HISTORIES

Unfortunately, experience and not foresight is typically the most convincing teacher. Case histories are illustrative of possible problems and how they were handled elsewhere. This chapter relates a number of case histories of problems with monitoring and recovery wells of various types and resolutions (if any) of these problems.

Environmental professionals sometimes view their sphere of influence as being unique. One problem in dealing with well deterioration in environmental systems is a lack of a sense of history — knowing what was done before. Knowing that thousands of water supply utilities and geotechnical and mining projects have dealt with encrustation, sanding, biofouling, and corrosion for over a century in modern times (we have only sketchy but intriguing descriptions of problems in ancient systems) must provide some comfort.

Recent Estimates of Well Deterioration Problem Frequencies in Wells

In a survey among AWWA-member utilities in the southern U.S. (Smith, 1990), iron biofouling, sand-pumping, hydrogen sulfide (H_2S), and corrosion were identified as the most predominant well maintenance problems. It is to be expected that biofouling is the root problem causing a large portion of the corrosion and hydrogen sulfide, as well as iron discoloration, but that the cause is undetected due to a lack of biofouling monitoring.

A parallel nationwide survey among water well contractors involved in municipal well rehabilitation (Smith, 1990) put "iron bacteria" at the top of reported problems, along with corrosion and H_2S, all reported by 20% of respondents.

Tests of wells over large areas by the author and others (e.g., Gehrels and Alford, 1990; Smith, 1992; Cullimore, 1993) have found biofouling organisms to be virtually ubiquitous in aquifer systems. The problems experienced by water supply systems must be considered to be the minimum that pumping wells on environmental projects will experience.

Some Water Supply Iron Biofouling Case Histories

Water supply system problems have been documented in a way that environmental system problems have not. There is of course more time for these systems to have been studied and more willingness to "go public." The following are two somewhat different examples. Several more useful case histories for both biofouling and other problems and solutions are summarized by Howsam (1990a,b), Smith (1992), and Borch, Smith, and Noble (1993). Several geotechnical and "environmental" case histories (mostly remediation) are provided next.

Public Water-Supply Well, Northwestern Ontario

Gehrels and Alford (1990) describe a situation in northwestern Ontario in which a water supply well for a small community exhibited a steady loss of efficiency, beginning 1 year after its construction in 1987. This well was constructed with a 0.050-in. (1.27-mm) slot stainless steel screen tapping a coarse, mixed outwash aquifer and initially was estimated to produce 17 l/s (269 gpm). From January 1988 to April 1989, the well yield decreased 65% from an already diminished level (13.85 to 6.06 l/s, 220 to 96 gpm), resulting in a noticeable water shortage.

Iron biofouling was suspected as a cause by the Ministry of Environment, and samples analyzed contained *Gallionella* and sulfate-reducing bacteria (SRB) (contributors to corrosion). Subsequent bacterial testing indicated that iron biofouling was severe.

The well was treated beginning in July 1989 using an iterative process of applying heat and disinfection chemicals (a version of the BCHT Process; see Chapter 9, Cullimore [1993] and Borch, Smith, and Noble [1993]). This resulted in an increase in yield to 11.5 l/s (182 gpm), but tests indicated the presence of a still-detectable microbial component. When the pump was pulled in August, numerous construction deficiencies were also noted. Their effects on the well problem were marginal, but certainly not helpful.

Yield subsequently decreased again until pretreatment levels were reached. A second round of more vigorous treatment resulted in a recovery of the original specific capacity. This performance has been maintained by an ongoing maintenance treatment program consisting of periodic shock chlorination and acidizing to control biofouling.

Lessons that can be learned: First, near-freezing groundwater temperatures are no barrier to severe iron biofouling problems. This contention can now be put to rest for good. A typical yield-reduction pattern for iron biofouling clogging is exhibited (initially not very noticeable, but accelerating and eventually dramatic). The treatment process employed was successful by most standards, but did not result in permanent improvement. Even an intensive rehabilitative treatment is not permanent and maintenance treatment is necessary for long-term operation. Biofouling testing detected residual iron-precipitating microflora, providing a warning of the need for additional treatment.

Public Water-Supply Ranney Wells, Belgrade, Serbia

Long before the breakup of Yugoslavia, Belgrade had other, quieter problems: iron biofouling plugging of Ranney (radial collector) and tube wells. Ranney collector wells per se are not a part of the environmental remediation culture, but problems with them have application in horizontal screened systems.

In Belgrade, extensive studies of the problem are described in a lengthy series of publications (Barbic et al., 1974; Barbic et al., 1975; Barbic et al., 1987; Barbic, Krajcic, and Savic, 1990). These studies extended over a period of 20 years up to the recent times, the longest such continuous studies known. Belgrade is supplied by approximately 52 Ranney wells and 8 tube wells approximately 30 m (100 ft) in depth in unconsolidated formations.

Ranney wells provide extensive contact with the producing aquifer due to their long, screened laterals, and in many cases apparently their hydraulic performance is less affected by massive microbial growth than vertical tube wells because their effective screen area is vastly larger. In the Belgrade case, however, the Ranney wells were experiencing severe yield reductions due to the plugging of the laterals. Iron bacteria were detected by microscopy in all groundwater samples, and iron biofouling as a whole seemed to increase with time if wells were not restored.

Lessons that can be learned: Iron biofouling can contribute to, if not solely cause, plugging and yield reduction even in a well design that provides extensive hydraulic contact with the aquifer. A combination of systematic maintenance monitoring and rehabilitation treatments serves to keep the wells functioning.

Geotechnical Case Histories

While an engineering dewatering situation is not the same as an aquifer remediation project, geotechnical engineering methods such as described in Powers (1992) are the prototypes for "environmental" pumping strategies. With regard to well operation, many of the problems are quite similar. The following are three case histories illustrative of well maintenance and rehabilitation concerns in dewatering projects.

Long-Term Dewatering, Eastern U.S.

This project is illustrative of problems with long-term pumping of organically rich and microbially active groundwater. This project was designed to lower water levels during some civil construction in sandy surficial aquifers with high water tables. Some of the groundwater was organically contaminated.

A system of submersible pumping wells and eductor systems was employed. The recirculation water of the eductor systems was treated with chemicals to limit clogging and biofouling. The construction took some months, and both the submersible and eductor wells experienced heavy iron deposition and apparent biofouling. The systems had to be disassembled and cleaned, delaying construction and causing some tension among the various parties involved (contractor, environmental engineering firm, and government entity).

Lessons to be learned: Pumping organically rich, microbially active groundwater for long periods results in extensive iron deposition in wells and related piping systems. Organics are microbially degraded and iron mobilized, then oxidized and deposited in the redox fringe around wells (see Chapter 2). A preventive maintenance treatment program was in place. However, a more effective program based on information available at the time in the open literature could have mitigated the problems encountered.

Biofouling of Site-Dewatering Systems, United Kingdom

While the above U.S. case history may seem like a case of freakish bad luck, the situations described by Powrie, Roberts, and Jefferis (1990) sound like they are written out of the same book. This paper also succinctly summarizes the operation and environment of construction dewatering systems.

Powrie, Roberts, and Jefferis (1990) summarize six cases:

Site A (alluvial sand and gravel, 200 l/min flow, high Fe) was pumped with submersibles that had to be pulled and cleaned at 2-month intervals. This was soft, easily removed biomass. High-chloride estuarine groundwater contributed to corrosion of the Type 304 stainless steel pump volutes.

Sites B and C (gravel over chalk, 330–500 l/min, low Fe) operated without serious interruption even though some biofilm developed. Site B pumps exhibited the Type 304 chloride corrosion observed at site A.

Sites D to F (silts and silty fine sand, 0.5–20 l/min, Fe moderately high) were pumped by ejector systems. The higher flow rate system (of two) at Site D experienced severe iron biofouling and interruption of operations during cleaning at 3- to 4-week intervals. The biofouling material remained soft and easily removed. Site E was similar, but was decommissioned before cleaning became necessary. Site F (low flow) exhibited biofilm development, but performance was not impacted.

Lessons to be learned: Problems are the same all over. Injection of biocides was ineffectual for the most part. Higher Fe levels in groundwater and higher flow rates resulted in more clogging. Systems operating under vacuum and anaerobically did not exhibit the clogging symptoms.

An Integral Well Cleaning System for Dam Pressure-Relief Wells, New South Wales, Australia

Iron biofouling is a common maintenance problem afflicting pressure relief wells reducing hydraulic head across earthen dams. The U.S. Army Engineers have a continuous rehabilitation effort on dams under their jurisdiction, currently preferring the BCHT Process (Chapter 9) for this work. Jewell (1990) describes a design process that anticipated iron biofouling problems incorporating (1) large screen open area to "buy time" until rehabilitation is necessary and to facilitate chemical contact with the aquifer, (2) corrosion-resistant materials, and (3) an integral cleaning system.

The cleaning system incorporates a chemical feed tube extending to the bottom of the well. Gas pressure can be applied to force chemicals out through the screen and purge water up the central feed tube.

Lessons to be learned: Well systems of many types can be designed to incorporate maintenance treatment features to control anticipated problems. Attention to maintenance concepts resulted in a system suitable for routine maintenance.

Environmental Site Case Histories

More so than in other technical topics (and more like Dear Abby), the identity and details of activities occurring on environmental management sites are extremely sensitive information. For that reason, factors that may serve to identify a site to anyone but those directly involved are obscured in these examples (except for those published), which may represent composites of types of problems from reports. They are actual situations. Paraphrasing *Dragnet,* the identities are concealed but the problems are real.

Pump-and-Treat Recovery Wells, Superfund Site, Glacial Sandy Site, Midwest U.S.

Pumping recovery wells were installed to recover a complex plume array including organics, metals, and radionuclides. Within 5 months, 2-in. discharge lines from the wells were reduced to 1/2 in., and flow was severely restricted by a soft iron-biofouling mass. The wells were mechanically cleaned, but the clogging quickly returned, reducing pipe diameters to 1/4 in.

Wells were then inspected by a professional well rehabilitation contractor and thoroughly cleaned using a multiple-step, multiple-chemical process and again returned to nominal performance. Upon recommendation of the contractor, flow was throttled back to reduce well aeration, and a preventive treatment program instituted, requiring treatment every 8 weeks with a mixture of hydroxyacetic acid and sulfamic acid and wetting agents. Eight weeks was determined to be within the timeframe when treatment was possible without full-scale well rehabilitation.

A contributing factor in the opinion of the contractor personnel present was poor design and construction of the pumping wells, so that aeration of the screened area and enhancement of fouling was encouraged. The wells were also overpumped due to the installation of pumps too large for the purpose. These problems have been addressed by site management.

Lesson that can be learned: Preventive design, material selection, and preventive maintenance, as well as improved operation could have mitigated the well problems. The treatment employed was successful under the circumstances when followed by preventive maintenance treatment.

Plume-Control Purge Wells, Great Lakes Area Landfills

Pumping wells are often installed to control existing plumes or to prevent groundwater flow from the direction of a landfill from reaching sensitive discharge points such as streams. In these two situations, wells were completed in upper sandy aquifers and pumped to discharge, which was collected and treated.

In both cases, overpumping and clogging of pumps and discharge lines by iron-fouling material became a problem. Examination of the clogging material and tests of pumped water revealed iron-related bacteria (IRB), SRB, and other biofouling bacteria. In the one case, numerous molds were also detected, a symptom of extensive dewatering of the aquifer down into the screen. The wells of the other site were not tested for molds. Design was implicated as a problem by contractor personnel in the one case, as screens were routinely dewatered and pumps cycled 14 or more times a day. Design and operation were less evident as operational problems in the other case, although the system design limited treatment options.

Final resolution of problems at both sites is still uncertain. Limited mechanical well and pump cleaning was initiated. Site management at both sites have not committed to full rehabilitation and maintenance programs several years after severe well performance problems were first identified.

Lesson that can be learned: Preventive design, material selection, and preventive maintenance, as well as improved operation, could have mitigated the well problems. Implementation of a recommended maintenance treatment program could keep them from getting worse, but this was not performed for unknown management reasons.

Product Recovery Wells, Great Lakes Area Industrial Spill Sites

Pumping wells and GAC systems designed and installed for recovery of distillates were fouled by slime-forming and S^{2-}-oxidizing bacteria (SOB), resulting in reduced well performance and rapid clogging of mechanical prefilters. The microbial fouling problem was identified and a treatment program recommended.

Chlorination was discarded as an option due to the need to contain and dispose of chlorinated organics. Since the well had a small volume and was near an industrial property, a hot-water recirculation system was recommended and installed by the environmental contractor. This system is used successfully to periodically treat the well.

Based on this experience, the environmental firm asked for a biofouling potential assessment of another well in a similar situation. Biofouling was indicated as a problem based on analyses, and preventive treatment measures instituted.

Lessons to be learned: Attention to microbial fouling potential could have prevented a nuisance problem, but willingness to correct a problem resulted in success. Lessons learned resulted in preventive measures on a second site.

Monitoring Well Deterioration and Rehabilitation, New England, U.S.

At a site in southern Maine, an array of monitoring wells had been installed over a period of time to monitor VOC groundwater contamination in a glacial till over fractured bedrock setting. The wells were screened in the till and the underlying rock. Erratic and questionable chemical data from the wells resulted in a decision to rehabilitate the wells. Surging was chosen as the rehabilitation method. A single-valve type surge was used, operated with a cathead.

During prerehabilitation testing, sedimentation was discovered in the well sumps. This material was apparently not affected during purging prior to sampling and may have been a part of the erratic data problem. VOCs partitioning on the sediment may have been released erratically. Reddish orange turbid water was removed during surging and "iron bacteria" were observed from surge samples taken from "the more contaminated wells at the site" in "small quantities." The iron coloration was attributed to the reaction between the TCE and wells constructed of black steel with stainless steel screens.

Rates of water level recovery after pumping improved after surging. Sump sediment was also removed. One apparent casing rupture was revealed, and the well was subsequently abandoned. Some apparent changes in TCE levels were observed, but the trends were inconclusive (Sevee and Maher, 1990).

Lessons to be learned from this published case history:

1. Reviewing the theoretical information provided in Chapters 1 and 2 of this work, the well material selection was inappropriate for a setting with high microbial corrosion potential.
2. The rehabilitative work revealed several conditions (apparent iron biofouling, sediment build-up, apparent casing breach) that could have been detected and corrected earlier if a regular maintenance plan was in place. However, until recently, there has been no doctrine of routine maintenance for such wells.
3. There was also no information presented on analyses of purge material to conclusively characterize them (biofouling or sediment, TCE content, etc.), or to pinpoint the nature of the well rupture reported. No microbial or physicochemical testing to identify corroding factors was reported. It is therefore not known if anything was learned about the well degrading conditions at the site. A follow-up report would be interesting.
4. The decisions made were astute and appropriate. Surging is an excellent redevelopment technique, which apparently provided good results despite the limited force that could be applied by the system described. A choice of chemicals probably would have improved results. The decision to abandon the ruptured well was also appropriate.
5. Questions remain. One question is whether the rehabilitation of wells not meeting modern monitoring well construction standards is worthwhile. If they can continue to provide acceptable data, probably so, at least for the short term. The second question is, will the data remain less

erratic? Continued maintenance should help, as long as the wells are not already too degraded.

Calcium Sulfate Plugging of Hydrocarbon-Recovery Well Pumps

An extensive array of hydrocarbon recovery wells at a site, designed to recover light free-phase hydrocarbon had experienced a high incidence of pump failure problems (Hodder and Peck, 1992). This problem began soon after the beginning of aquifer remediation efforts in one area. Pump failure occurred reliably after 1 week to 1 month of operation, but proceeded gradually rather than suddenly. The frequent pump failures were hampering efforts to increase pumping and to clean up the site.

Dismantled pumps were found to be clogged with a reddish-brownish to white substance, primarily consisting of $CaSO_4$ with NaCl, various iron oxides or hydroxides, and bacterial colonies including SOB. Some wells had evidence of iron biofouling.

Tackling the problem involved several stages. Different pump designs were tried with marginal to negative results. A particularly robust design facilitated pump rebuilding, but did not improve service life. A solution of 20% hydrochloric acid was tried on the clogging material, resulting in rapid dissolution of much of the material. Some dissolved slowly and about 1% black material did not dissolve.

An acid fluid purge apparatus was devised and its use implemented in a preventive treatment program after a pilot test. Acid delivery was made to the pump initially when discharge pressure reduced; but as this became predictable, acid was fed preventively (10 to 40 gal per treatment).

Results:

1. Impeller wear continued to be a problem; however, pumps could be kept operating between 3 and 6 months. Treated pumps inspected were generally free of attached precipitant. However, Teflon bearings and seals were worn due to pumping loose abrasive precipitant. New seal types are being tried.
2. Water flow could be maintained and hydrocarbon recovery increased by 20%.
3. Well maintenance costs were reduced.
4. Downstream piping and treatment was unaffected. The high overall flow (>1500 gpm) and the neutralizing effects of the sedimentation were the reason.
5. An effective and safe portable treatment system was designed and employed.

Lessons to be learned from this published case history:

1. Well fouling problems can have a major negative impact on aquifer remediation project goals.

2. An analysis of the groundwater biogeochemistry from pilot pumping test wells could have revealed the need for a preventive maintenance treatment regime. The nature of this remediation and the aquifer environment that led to the pump problems are not unusual.
3. Even a limited maintenance program implemented after the fact can have a positive result for the entire project.
4. A program of identifying the problem, testing a possible solution, and implementing the solution is the way to solve this type of operational problem. The project personnel were able to identify a cause, develop a treatment, and see a result from applying a treatment.
5. Refinement continues. Once initial problems and cures are identified, loose ends can be tackled, for example, the bearing problems. Further testing may reveal the root causes of the $CaSO_4$ precipitate (likely secondary mineralization of the products of a selection of SOB) resulting in well modifications that may retard their growth, and improved treatment chemical selection and application.

Jet Fuel Spill Remediation Project, Air Force Base, Eastern U.S.

Wells were constructed to remove jet fuel and contaminated groundwater from a silty surficial Coastal Plain aquifer (about 0.5 million gal spilled over many years). In the initial phase, several wells were constructed (6-in. diameter and about 60 ft deep with 25- to 40-ft PVC screens). The exact as-built dimensions are unknown because a former contractor apparently removed records from the site (not unusual, unfortunately).

These wells plugged with an apparent biofouling, aggravated by silting in about 1 year. They were mechanically redeveloped and clogged again in 6 to 7 months (a typical progression). Submersible pumps also clogged. These were found to be not optimal for the application (inefficient at the operational pumping rate). The wells were then rehabilitated with a more vigorous staged chemical and agitation treatment, pumps refurbished, and the system placed on a maintenance treatment program that is yet to be implemented as of this writing. The contractor has recommended well and pump improvements for the next round of construction planned.

Lessons to be learned: Proper development during construction would have bought some time until clogging occurred. More appropriate pump selection may also have helped, but there is a limited selection of pumps available. A staged chemical well rehabilitation treatment with vigorous surging was useful in mitigating clogging, but maintenance treatments will be necessary for the long term.

Contaminated Groundwater Extraction Wells, New England, U.S.

Wells vulnerable to iron biofouling clogging on more than one site had been either cleaned or projected to biofoul at some time in the future. Maintenance programs based on maintenance monitoring and regular maintenance treatment and pump removal and cleaning is keeping wells open and operating. Some system engineering that stood in the way of maintenance are being corrected.

Lessons to be learned: Preventive maintenance works, and site consultants and engineers willing to incorporate well and system maintenance are having fewer problems with their systems.

WHERE DO WE GO FROM HERE?

There is a lot to be learned on the practical level. The root causes of the types of well problems occurring are probably well enough understood in a theoretical way at this moment in history to develop useful preventive maintenance and well rehabilitation action plans. The intrinsic value of prevention is also well demonstrated, but the gospel message has yet to be well spread through the environmental engineering empire.

Wish List

There are a number of definite needs and wants in environmental well maintenance:

1. *The useful maintenance monitoring methods available now need to be used systematically.* Further refinement in monitoring methods would be helpful in providing earlier detection and reliable quantitative data on well and aquifer biogeochemistry. More research and site-level experience is needed with monitoring methods to provide the desired refinement, and the usefulness of standardization of methods is an open question, but using the methods we have now will provide useful early warning maintenance monitoring.
2. *Controlled research into more effective and, at the same time, environmentally benign preventive and restorative treatments.* There are a lot of candidate methods and continual refinement, but controlled and unbiased research into their efficacy is needed. Remarkably, this has never been done. It is needed so that methods proposed for specific jobs can be objectively evaluated.
3. *The economics of well maintenance have to be updated in general and specifically for environmental projects.* No industry-wide information is available, only case histories. There is nothing in modern literature comparable to studies published in 1960 and 1961 by L. Koenig (summarized in Borch, Smith, and Noble, 1993) for water supply wells (in an economy that is now ancient history).
4. *A more extensive documented body of experience with well maintenance on monitoring and remediation wells is badly needed.* Right now what is out there is limited in value because the case histories available involve work that is patchy, short in duration, and hampered by the need for site confidentiality.

5. *Above all, what is needed is an appreciation for a preventive mainte-nance ethos, the complexity of the causes of well deterioration, and how it can be prevented and tackled in a targeted way.* Environmental pump-ing and monitoring wells (and wells in general) need to be maintained the way that other expensive, long life-cycle structures are maintained. *People need to appreciate that well maintenance is most comparable to medicine (and maybe modern auto service) at the present time:*
   a. Maintenance is less costly than rehabilitation.
   b. Lack of maintenance will kill you over the long haul.
   c. The causes of deterioration are complex.
   d. Such maintenance cannot be planned without some expert involve-ment and adjustment over time.

People in site management have to be practical and open-minded about this, with a view to the long-term, not the immediate, situation. We also need a range of (we hope) ever-improving maintenance options available.

Education, Communication, and Mutual Respect: Human Issues in Environmental Well Maintenance

Experience shows that people in the trenches of environmental remediation work find well maintenance and rehabilitation to be frustrating and unglamorous, and the solutions intimidating. There are no quick and easy solutions; but with some communication, perhaps the frustration and intimidation can be alleviated. *(Unglamorous it remains.)*

There are two somewhat opposing tendencies determining the number of options available for well maintenance and rehabilitation in the environmental industries: (1) on the one hand are the increasing number of products available and the inventiveness of well rehabilitation practitioners; and (2) on the other hand, there is pressure to disapprove any and all chemical use in the maintenance of groundwater systems. This is like washing when you are really dirty, *without soap and detergent.*

No reasonable person wants to contaminate the biosphere. People in respon-sible charge should have a firm understanding of the consequences of their actions. Level heads and education are needed to keep a reasonable arsenal of treatments available to keep the wells operating. *If they cannot operate, they will be incapable of reliably providing samples to detect contaminants or extracting water to clean it up or keep it away from sensitive environments.*

Achieving this balance takes broad training and communication, mutual respect, and team approach. Site management, hydrogeologists, engineers, drill-ing and pump technicians, each must value the skills, experience, and advice of the other.

All of this is necessary: the scientific and engineering principles, and the empirical applied experience that knows how pumps and wells work and fail in the well, and what to do about it (and when to quit).

Site management and consultants can start by keeping a healthy humility in the face of and respect for the value of experience and specialized training in aspects of well M&R. Saying this as delicately as possible, sometimes such humility and respect does not seem to be natural to many educated people these days, and must be learned by example in company culture, since it is not taught in higher education environments.

On a remedial education level, the message that has to be heard through the symposia and short-course circuit, and in company and agency training throughout the industrial world, includes:

1. Operating monitoring and remediation wells is high maintenance. The potential for problems is great, but manageable.
2. Design preventively and for the long term — not quick and cheap — no matter what the pressures.
3. Plan rationally for maintenance and rehabilitation.
4. Base design and maintenance on adequate diagnosis of potential problems (which may be numerous and interactive).
5. If you have a problem, get help and fix it properly as quickly as possible — don't wait or apply "half-baked" solutions.

There have to be lines of communication among all the payers and players in the groundwater and environmental remediation industries. Project management and site supervision needs to be listening to the people who have made the study of well deterioration problems and solutions their specialties. Doing so will save a lot of grief and expense.

We all need to talk to one another and share our experiences as much as possible. Echoing Roy Cullimore's sentiments in his recent monograph (Cullimore, 1993), it is hoped that this work should also only be part of the beginning of a dialog on the maintenance of groundwater monitoring, remediation, recovery, and plume control wells. Indeed, it may even become obsolete as new methods and experience changes what we know and recommend (that's for the next edition). If we talk these problems out, write up our successes and failures, conduct and fund the needed research, share our experiences and expertise, while keeping specialists in this field gainfully employed doing what they do best (instead of selling water filters or something), we can keep these projects operating and serving their purpose: controlling groundwater contamination.

References

The following references were used in preparation of this work. They are recommended to the reader for more detailed study of specific topics and situations. Some useful appendix material is also included for reference.

Alcalde, R. E. and M. A. Gariboglio. 1990. Biofouling in Sierra Colorado water supply: A case study, in *Microbiology in Civil Engineering*. P. Howsam, Ed. FEMS Symposium No. 59, E.&F.N. Spon, London, pp. 183-191.

Alford, G., N. Mansuy, and D. R. Cullimore. 1989. The utilization of the blended chemical heat treatment (BCHT) process to restore production capacities to biofouled water wells, in *Proc. of the Third Annu. Outdoor Action Conf., Las Vegas, NV*. National Water Well Association, Dublin, OH, pp. 229-237.

Aller, L. et al. 1990. *Handbook of Suggested Practices for the Design and Installation of Ground-Water Monitoring Wells*. National Water Well Association, Dublin, OH.

Alper, J. 1992. Microbial alchemy: Golden rods and cocci. *ASM News* 58(12):650-651.

APHA, AWWA, WEF. 1992. Section 9240 — Iron and Sulfur Bacteria, in *Standard Methods for the Examination of Water and Wastewater*. American Public Health Association, Washington, D.C., pp. 9-73 to 9-83.

Atlas, R. M. 1993. *Handbook of Microbiological Media*. CRC Press, Boca Raton, FL.

Bakke, R. and P. O. Olsson. 1986. Biofilm thickness measurements by light microscopy. *J. Microbiol. Meth.* 5:93-98.

Barbic, F. F. et al. 1974. Iron and manganese bacteria in Ranney wells. *Water Res.* 8:895-898.

Barbic, F .F. et al. 1975. Development of iron and manganese bacteria in Ranney wells. *J. AWWA* 67:565-572.

Barbic, F. F. et al. 1987. Ecology of iron and manganese bacteria in underground water, in *Proc. 1986 Int. Symp. Biofouled Aquifers: Prevention and Restoration*. D. R. Cullimore, Ed. American Water Resources Assn., Bethesda, MD, pp. 11-22.

Barbic, F. F., O. Krajcic, and I. Savic. 1990. Complexity of causes of well yield decrease, in *Microbiology in Civil Engineering*. P. Howsam, Ed. FEMS Symposium No. 59, E.&F.N. Spon, London, pp. 198-208.

Beller, H. R., D. Grbic-Galic, and M. Reinhard. 1992. Microbial degradation of toluene under sulfate-reducing conditions and the influence of iron on the process. *Appl. Environ. Microbiol.* 58:786-793.

Borch, M. A., S. A. Smith, and L. N. Noble. 1993. *Evaluation, Maintenance, and Restoration of Water Supply Wells*. American Water Works Association Research Foundation, Denver, CO.

Chapelle, F. H. 1992. *Ground-Water Microbiology and Geochemistry*. John Wiley & Sons, New York.

Chapelle, F. H. and D. R. Lovley. 1992. Competitive exclusion of sulfate-reduction by Fe(III)-reducing bacteria: A mechanism for producing discrete zones of high-iron ground water. *Ground Water* 30(1):29-36.

Characklis, W. G. et al. 1987. Biofilms in porous media, in *Proc. 1986 Int. Symp. Biofouled Aquifers: Prevention and Restoration*. D. R. Cullimore, Ed. American Water Resources Assn., Bethesda, MD, pp. 57-78.

Christian, R. D. 1975. Distribution, Cultivation, and Chemical Destruction of *Gallionella* from Alabama Ground Water. M.S. Thesis. The University of Alabama, Tuscaloosa, AL.

Corstjens, P. L. A. M. et al. 1992. Enzymatic iron oxidation by *Leptothrix discophora*: Identification of an iron-oxidizing enzyme. *Appl. Environ. Microbiol.* 58:450-454.

Cote, R. J. and R. L. Gherna. 1994. Nutrition and media (Chapter 7), in *Methods for General and Molecular Bacteriology*, American Society for Microbiology, Washington, D.C., pp. 155-178.

Craig, M. 1991. Developing monitoring wells. *Ground Water Age* 25(10):16-19.

Cullimore, D. R. 1981. The Bulyea experiment. *Canadian Water Well* 7(3):18-21.

Cullimore, D. R. 1993. *Practical Ground Water Microbiology*. Lewis Publishers, Chelsea, MI.

Cullimore, D. R. and A. E. McCann. 1977. The identification, cultivation, and control of iron bacteria in ground water. In *Aquatic Microbiology*. F. A. Skinner and J. M. Shewan, Eds. New York, Academic Press, pp. 219-261.

Davis, J. A. et al. 1993. Influence of redox environment and aqueous speciation on metal transport in groundwater: Preliminary results of trace injection studies, in *Metals in Groundwater*. H. E. Allen et al., Eds. Lewis Publishers, Chelsea, MI, pp. 223-273.

Driscoll, F. G. 1986. *Groundwater and Wells*. Johnson Division, St. Paul, MN.

Edil, T. B. et al. 1992. Sealing characteristics of selected grouts for water wells. *Ground Water* 30:351-361.

Eggington, H. F. et al. 1992. *Australian Drilling Manual*. Australian Drilling Industry Training Committee Ltd., Macquarie Centre, NSW.

Ehrhardt, G. and R. Pelzer. 1992. Wirkung von Saugstromsteuerungen in Brunnen. *bbr* Okt. 1992, 43, Jahrgang, pp. 452-458. (English translation, "Effect of suction flow control devices in wells," available from the author of this text.)

Emerson, D. and W. C. Ghiorse. 1992. Isolation, cultural maintenance, and taxonomy of a sheath-forming strain of *Leptothrix discophora* and characterization of manganese-oxidizing activity associated with the sheath. *Appl. Environ. Microbiol.* 58:4001-4010.

Fish, W. 1993. Subsurface redox chemistry: A comparison of equilibrium and reaction-based approaches, in *Metals in Groundwater*. H. E. Allen et al., Eds. Lewis Publishers, Chelsea, MI, pp. 73-101.

Fountain, J. and P. Howsam. 1990. The use of high-pressure water jetting as a rehabilitation technique, in *Water Wells Monitoring, Maintenance, and Rehabilitation*. P. Howsam, Ed. E.&F.N. Spon, London, pp. 180-194.

Fredrickson, J. K. et al. 1989. Lithotrophic and heterotrophic bacteria in deep subsurface sediments and their relations to sediment properties. *Geomicrobiol. J.* 7(1-2):53-66.

Gaber, M. S. and B. O. Fisher. 1988. *Michigan Water Well Grouting Manual*. Michigan Dept. of Health, Lansing, MI.

Gäss, T. E. et al. 1991. Test results of the Grundfos ground-water sampling pump, in *Proc. of the Fifth Natl. Outdoor Action Conf., Las Vegas, NV*. National Ground Water Association, Dublin, OH.

Gehrels, J. and G. Alford. 1990. Application of physico-chemical treatment techniques to a severely biofouled community well in Ontario, Canada, in *Water Wells Monitoring, Maintenance, and Rehabilitation*. P. Howsam, Ed. E.&F.N. Spon, London, pp. 219-235.

Gottfreund, E., J. Gottfreund, and R. Schweisfurth. 1985. Occurrence and activities of bacteria in the saturated and unsaturated underground in relation to the removal of iron and manganese. *Water Supply* 3 (Berlin A):109-115.

Gounot and di Ruggiero. 1990. Reduction and migration of manganese in groundwaters (poster). Microbiology in Civil Engineering, Cranfield Institute of Technology, Cranfield, U.K., September 1990.

Grundfos Pumps Corp. 1992a. Installation and operating instructions, Grundfos Redi-Flo4 stainless steel submersible pumps for environmental applications (circular). Clovis, CA.

Grundfos Pumps Corp. 1992b. Redi-Flo 2 installation and operating instructions (circular). Clovis, CA.

Hacket, G. and J. H. Lehr. 1985. *Iron Bacteria Occurrence, Problems and Control Methods in Water Wells*. National Water Well Association, Dublin, OH.

Hallbeck, L. 1993. On the biology of the iron-oxidizing and stalk-forming bacterium *Gallionella ferruginea*. Department of General and Marine Microbiology, University of Göteborg, Göteborg, Sweden.

Hallbeck, E.-V. and K. Pedersen. 1987. The biology of *Gallionella*, in *Proc. 1986 Int. Symp. Biofouled Aquifers: Prevention and Restoration*. D. R. Cullimore, Ed. American Water Resources Assn., Bethesda, MD, pp. 87-95.

Hallberg, R. O. and R. Martinell. 1976. VYREDOX™ — In situ purification of ground water. *Ground Water* 14:88-93.

Hanert, H. H. 1981. The genus *Gallionella*, in *The Procaryotes: A Handbook on Habitats, Isolation, and Identification of Bacteria*. M. P. Starr et al., Eds. Springer-Verlag, Berlin, pp. 509-521.

Hargesheimer, E. E., C. M. Lewis, and C. M. Yentsch. 1992. *Evaluation of Particle Counting as a Measure of Treatment Plant Performance*. AWWA Research Foundation, Denver, CO.

Hässelbarth, U. and D. Lüdemann. 1972. Biological incrustation of wells due to mass development of iron and manganese bacteria. *Water Treat. Exam.* 21:20-29.

Helweg, O. J., V. H. Scott, and W. C. Scalmanini. 1983. *Improving Well and Pump Efficiency*. American Water Works Association, Denver, CO.

Hem, J. D. 1985. *Study and Interpretation of the Chemical Characteristics of Natural Water*, 3rd ed. Water Supply Paper 2254, U.S. Geological Survey, Reston, VA.

Hodder, E. A. and C. A. Peck. 1992. Aquifer restoration system improvement using an acid fluid purge, in *Proc. of the Sixth National Outdoor Action Conference, Las Vegas, NV*. National Ground Water Association, Dublin, OH, pp. 471-482.

Holt, J. G. and N. R. Krieg. 1994. Enrichment and isolation (Chapter 8), in *Methods for General and Molecular Bacteriology*. American Society for Microbiology, Washington, D.C., pp. 179-215.

Howsam, P., Ed., 1990a. *Microbiology in Civil Engineering*. FEMS Symposium No. 59, E.&F.N. Spon, London.

Howsam, P., Ed. 1990b. Water wells monitoring, maintenance, and rehabilitation, in *Proc. of the Int. Groundwater Engineering Conf.* Cranfield Institute of Technology, UK. E.&F.N. Spon, London.

Howsam, P., B. Misstear, and C. Jones. 1994. *Monitoring, Maintenance and Rehabilitation of Water Supply Boreholes.* Construction Industry Research and Information Association, London, England.

Howsam, P. and S. Tyrrel. 1989. Diagnosis and monitoring of biofouling in enclosed flow systems — Experience in groundwater systems. *Biofouling* 1:343-351.

Howsam, P. and S. Tyrrel. 1990. *An International Survey of Attitudes and Circumstances Relating to the Monitoring, Maintenance, and Rehabilitation of Water Supply Boreholes.* Report to the Overseas Development Administration (Report 4582), Silsoe College, Silsoe, England.

Interagency. 1985. *Protecting Health and Safety at Hazardous Waste Sites: An Overview* (EPA/625/9-85-006). U.S. Environmental Protection Agency, Washington, D.C.

Jewell, C. M. 1990. Design of pressure relief wells with an integral cleaning system for a large earth-fill dam, in *Microbiology in Civil Engineering.* P. Howsam, Ed. FEMS Symposium No. 59, E.&F.N. Spon, London, pp. 317-327.

Kill, D. L. 1990. Monitoring well development — Why and how, in *Ground Water and Vadose Zone Monitoring.* STP 1053. D. M. Nielsen and A. I. Johnson, Eds. American Society for Testing and Materials, Philadelphia, PA, pp. 82-90.

Kraemer, C. A., J. A. Schultz, and J. W. Ashley. 1991. Monitoring well post-installation considerations, in *Practical Handbook of Ground-Water Monitoring.* D. M. Nielsen, Ed. Lewis Publishers, Chelsea, MI, pp. 333-365.

Leach, R., A. Mikell, C. Richardson, and G. Alford. 1991. Rehabilitation of Monitoring, Production, and Recharge Wells. *Proc. 15th Annual Army Environmental R&D Symp. (1990).* CETHA-TS-CR-91077, U.S. Army Toxic and Hazardous Materials Agency, Aberdeen Proving Grounds, MD, pp. 623-646.

LeChevallier, M. W. and W. D. Norton. 1992. Examining relationships between particle counts and *Giardia, Cryptosporidium,* and turbidity. *J. AWWA* 84(12):54-60.

Lehr, J. H. 1985. Where the sand goes: or when will the other shoe fall? *Water Well J.* 39(10):48-50.

Lewis, C. M., E. E. Hargesheimer, and C. M. Yentsch. 1992. Selecting particle counters for process monitoring. *J. AWWA* 84(12):46-51.

Lovley, D. R. 1991. Dissimilatory Fe(III) and Mn(IV) reduction. *Microbiol. Rev.* 55:259-287.

Lovley, D. R. and D. J. Lonergan. 1990. Anaerobic oxidation of toluene, phenol, and *p*-cresol by the dissimilatory iron-reducing organism, GS-15. *Appl. Environ. Microbiol.* 56:1858-1864.

Lutenegger, A. J. and D. J. DeGroot. 1994. Hydrologic properties of contaminant transport barriers as borehole sealants, in *Hydraulic Conductivity and Waste Contaminant Transport in Soils.* ASTM STP 1142. D. E. Daniel and S. J. Trautwein, Eds. American Society for Testing and Materials, Philadelphia, PA.

Macaulay, D. 1988. Maintenance for monitoring wells. *Ground Water Age* 22(8):24-27.

Mallard, G. E. 1981. Effects of bacteria on the chemical and physical state of iron, in *Microbiology of the Aquatic Environment.* Geological Survey Circular 848-E. U.S. Geological Survey, Reston, VA, pp. E13-E21.

Mansuy, N., C. Nuzman, and D. R. Cullimore. 1990. Well problem identification and its importance in well rehabilitation, in *Water Wells Monitoring, Maintenance, and Rehabilitation.* P. Howsam, Ed. E.&F.N. Spon, London, pp. 87-99.

Mouchet, P. 1992. From conventional to biological removal of iron and manganese in France. *J. AWWA* 84(4):158-167.

Nielsen, D. M., Ed. 1991. *Practical Handbook of Ground-Water Monitoring.* Lewis Publishers, Chelsea, MI.

Nielsen, D. M. and R. Schalla. 1991. Design and installation of ground-water monitoring wells, in *Practical Handbook of Ground-Water Monitoring*. D. M. Nielsen, Ed. Lewis Publishers, Chelsea, MI, pp. 239-331.

NIOSH. 1985. *Occupational Safety and Health Guidance Manual for Hazardous Waste Site Activities*, DHHS/NIOSH 85-115. National Institute for Occupational Health and Safety, Cincinnati, OH.

NIOSH. 1990. *NIOSH Pocket Guide to Chemical Hazards*. DHHS/NIOSH 90-117. National Institute for Occupational Health and Safety, Cincinnati, OH.

Norton, G. 1992. Pump service at hazardous sites. *Ground Water Age* 26(8):14-15.

Nuckols, T. E. 1990. Development of small diameter wells, in *Proc. Fourth Natl. Outdoor Action Conf. Aquifer Restoration, Ground Water Monitoring and Geophysical Methods*. National Water Well Assn., Dublin, OH, pp. 193-207.

Nuzman, C. E. and R. C. Jackson. 1990. Aquastream suction flow control device. *Proc. Conserv. 90, The National Conf. and Exposition, August 12-16, 1990, Phoenix, AZ*. National Water Well Assn., Dublin, OH, pp. 1275-1276.

Pedersen, K. 1982. Method for studying microbial biofouling in flowing-water systems. *Appl. Environ. Microbiol.* 43:6-13.

Pelzer, R. and S. A. Smith. 1990. Eucastream suction flow control device: An element for optimization of flow conditions in wells, in *Water Wells Monitoring, Maintenance, and Rehabilitation*. P. Howsam, Ed. E.&F.N. Spon, London, pp. 209-216.

Pope, D. H. et al. 1989. Microbiological aspects of microbiologically influenced corrosion, in *Microbial Corrosion: 1988 Workshop Proc.* ER-6345, Electrical Power Research Institute, Palo Alto, CA, pp. 3-1 to 3-24.

Powers, J. P. 1992. *Construction Dewatering*. Wiley-Interscience, New York.

Powrie, W., T. O. L. Roberts, and S. A. Jefferis. 1990. Biofouling of site dewatering systems, in *Microbiology in Civil Engineering*. P. Howsam, Ed. FEMS Symposium No. 59, E.&F.N. Spon, London, pp. 341-352.

Rich, C. A. and B. M. Beck. 1990. Experimental screen design for more sediment-free sampling, in *Ground Water and Vadose Zone Monitoring*. STP 1053. D. M. Nielsen and A. I. Johnson, Eds. American Society for Testing and Materials, Philadelphia, PA, pp. 76-81.

Riss, A. and R. Schweisfurth. 1985. Basic investigation about denitrification and nitrate-ammonification during the degradation of organic pollutions in the underground. *Water Supply* 3 (Berlin B):27-34.

Roscoe Moss Co. 1990. *Handbook of Ground Water Development*. John Wiley & Sons, New York.

Schalla, R. and W. H. Walters. 1990. Rationale for design of monitoring well screens and filter packs, in *Ground Water and Vadose Zone Monitoring*. STP 1053. D. M. Nielsen and A. I. Johnson, Eds. American Society for Testing and Materials, Philadelphia, PA, pp. 64-75.

Seal, K. J. 1990. Biodeterioration of materials used in civil engineering, in *Microbiology in Civil Engineering*. P. Howsam, Ed. FEMS Symposium No. 59, E.&F.N. Spon, London, pp. 39-52.

Sevee, J. E. and P. M. Maher. 1990. Monitoring well rehabilitation using the surge block technique, in *Ground Water and Vadose Zone Monitoring*. STP 1053. D. M. Nielsen and A. I. Johnson, Eds. American Society for Testing and Materials, Philadelphia, PA, pp. 91-97.

Smith, S. A. 1989. *Manual of Hydraulic Fracturing Methods for Well Stimulation and Geologic Studies:* National Water Well Association, Dublin, OH.

Smith, S. A. 1990. Well maintenance and rehabilitation in North America: An overview, in *Water Wells Monitoring, Maintenance, and Rehabilitation.* P. Howsam, Ed. E.&F.N. Spon, London, pp. 8-16.

Smith, S. A. 1991. Microbiological characterization of selected northwest Ohio carbonate-aquifer wells (abstract no. 20616). *GSA Abstracts with Programs* 23(3):61.

Smith, S. A. 1992. *Methods for Monitoring Iron and Manganese Biofouling in Water Supply Wells.* AWWA Research Foundation, Denver, CO.

Smith, S. A. and O. H. Tuovinen. 1990. Biofouling monitoring methods for preventive maintenance of water wells, in *Water Wells Monitoring, Maintenance, and Rehabilitation.* P. Howsam, Ed. E.&F.N. Spon, London, pp. 75-81.

Stout, J. E., V. L. Yu, and M. E. Best. 1985. Ecology of *Legionella pneumophila* within water distribution systems. *Appl. Environ. Microbiol.* 49:221-228.

Sutherland, D. C., P. Howsam, and J. Morris. 1993. *The Cost-Effectiveness of Monitoring and Maintenance Strategies Associated with Groundwater Abstraction — A Methodology for Evaluation.* ODA Project Report 5478A. Silsoe College, Silsoe, Bedford, U.K.

Tiller, A. K. 1990. Biocorrosion in civil engineering, in *Microbiology in Civil Engineering.* P. Howsam, Ed. FEMS Symposium No. 59, E.&F.N. Spon, London, pp. 24-38.

Tobiason, J. E. et al. 1992. Pilot study on the effect of ozone and PEROXONE on in-line direct filtration. *J. AWWA* 84(12):72-85.

Tuhela, L., S. A. Smith, and O. H. Tuovinen. 1993. Flow-cell apparatus for monitoring iron biofouling in water wells. *Ground Water* 31:982-988.

Van Loosdrecht, M. C. M. et al. 1990. Influence of interfaces on microbial activity. *Microbiol. Rev.* 54:75-87.

Van Riemsdijk, W. H. and T. Hiemstra. 1993. Adsorption to heterogeneous surfaces, in *Metals in Groundwater.* H. E. Allen et al., Eds. Lewis Publishers, Chelsea, MI, pp. 1-36.

Vuorinen, A. and L. Carlson. 1985. Scavenging of heavy metals by hydrous Fe and Mn oxides precipitating from groundwater in Finland, in *Proc. Int. Conf. Heavy Metals in the Environment,* 10-13 Sept., Athens, Greece, Vol. 1, T. D. Lekkas, Ed. University of Helsinki, Helsinki, Finland, pp. 266-268.

Waite, T. D. and T. E. Payne. 1993. Uranium transport in the subsurface environment, Koongarra — A case study, in *Metals in Groundwater.* H. E. Allen et al., Eds. Lewis Publishers, Chelsea, MI, pp. 349-410.

Water Systems Council. 1992. *Large Submersible Pumps Manual.* Chicago, IL.

Winegardner, D. L. 1990. Monitoring wells: Maintenance, rehabilitation, and abandonment, in *Ground Water and Vadose Zone Monitoring.* STP 1053. D. M. Nielsen and A. I. Johnson, Eds. American Society for Testing and Materials, Philadelphia, PA, pp. 98-107.

Wolfe, R. S. 1958. Cultivation, morphology, and classification of iron bacteria. *J. AWWA* 42:849-858.

ASTM Standards:

D 932-85, Iron Bacteria. 1991 Annual Book of Standards, Vol. 11.02, Water II, p. 491.

D 3977-80, Suspended sediment in water samples. 1991 Annual Book of Standards, Vol. 11.02, Water II, p. 674.

D 5088-90, Practice for decontamination of field equipment used at nonradioactive waste sites. 1993 Annual Book of Standards, Vol. 4.08, P. 1207.

D 5092-90, Vol. 4.08, p. 1210, Practice for design and installation of ground water monitoring wells in aquifers. 1993 Annual Book of Standards, Vol. 4.08, P. 1210.

D 5521-93, Standard guide for development of ground-water monitoring wells in granular aquifers, in press (ASTM Annual Book of Standards, Vol. 4.08, but did not make 1993 or 1994 editions).

Appendix: Method Selection Charts

The following appendix charts are provided for additional general reference.

Table A.1 Prevention and Mitigation Methods and Strategies

Problems to be prevented or controlled

Strategies and Actions	Pumping water level (s) decline	Reduced or insufficient yield (Q)	Lower specific capacity (Q/s)	Loss of or no production	Sand/silt/clay infiltration	Water quality changes	Pump corrosion and wear	Chemical incrustation or corrosion	Biofouling and microbial corrosion	Well structural failure	Loss of pump function
Efficient well design	◆	◆	◆					◆	◆		
Structurally sound materials and design				◆	◆	◆	◆			◆	◆
Choose corrosion-resistent materials based on water quality analysis	◆	◆	◆		◆	◆	◆	◆	◆	◆	◆
Quality drilling, construction, development	◆	◆	◆	◆	◆		◆		◆	◆	
Proper pump selection & installation		◆	◆	◆		◆	◆				◆
Appropriate hydrologic well operation	◆	◆	◆	◆	◆	◆	◆			◆	
Regular maintenance monitoring Preventive treatments (as appropr.)	◆	◆	◆				◆		◆		
Mechanical preventive maintenance		◆	◆	◆	◆	◆	◆				◆
Organized maintenance records program	◆	◆	◆	◆	◆	◆	◆	◆	◆	◆	◆
Professionally developed Maintenance plan	◆	◆	◆	◆	◆	◆	◆	◆	◆	◆	◆

Table A.2 *Diagnostic Methods Selection*

***Problems* to be prevented or controlled**

Diagnostic methods	Pumping water level (s) decline	Reduced or insufficient yield (Q)	Lower specific capacity (Q/s)	Loss of or no production	Sand/silt/clay infiltration	Water quality changes	Pump corrosion and wear	Chemical incrustation or corrosion	Biofouling and microbial corrosion	Well structural failure	Loss of pump function
Pumping or slug test (well/aquifer)	◆	◆	◆		◆	◆		◆	◆		
Pump test (pump operation)			◆	◆			◆	◆	◆		◆
Review well/pump design & construction, check operational records	◆	◆	◆	◆	◆	◆	◆	◆	◆	◆	◆
Check for turbidity, sand, silt	◆	◆	◆	◆	◆	◆	◆			◆	◆
Analyze for biofouling & physicochemical parameters of maintenance interest	◆	◆	◆			◆	◆	◆	◆		
Geophysical logs as appropriate (TV, caliper, gamma, gamma-gamma, acoustic)			◆			◆		◆	◆	◆	
Compare data to well construction and materials tolerances (corrosion, stress)				◆	◆	◆		◆	◆	◆	◆
Pull/inspect pump and well components	◆	◆	◆	◆	◆		◆	◆	◆	◆	◆
Analyze site/regional geologic & hydrologic information	◆	◆	◆	◆	◆	◆			◆		◆
Before proceeding, do homework: literature, suppliers, experts	◆	◆	◆	◆	◆	◆	◆	◆	◆	◆	◆

Table A.3 *Rehabilitation Methods Selection*

Problems to be prevented or controlled

Rehabilitation actions and methods	Pumping water level (s) decline	Reduced or insufficient yield (Q)	Lower specific capacity (Q/s)	Loss of or no production	Sand/silt/clay infiltration	Water quality changes	Pump corrosion and wear	Chemical incrustation or corrosion	Biofouling and microbial corrosion	Well structural failure	Loss of pump function
Is the rehabilitation method appropriate for site activities and goals?	♦		♦			♦	♦	♦	♦		
Review well/pump design & construction, hydrogeologic information	♦	♦	♦	♦	♦	♦	♦	♦	♦	♦	♦
Develop a plan and specifications that can guide work to meet goals		♦	♦	♦	♦	♦	♦	♦	♦	♦	♦
Replace/repair pump, well components		♦	♦	♦	♦		♦	♦	♦	♦	♦
Well redevelopment (appropriate selection)	♦	♦	♦		♦			♦	♦		
Agitation with chemicals and heat, CO₂ (appropriate selection, use)	♦		♦			♦		♦	♦		
Install/replace screen, pack, SFCD, sand separator					♦			♦	♦		
Well reconstruction or decommissioning as appropriate	♦	♦		♦	♦	♦				♦	
Install, renew flow and level monitoring equipment and begin/renew preventive maintenance	♦		♦	♦		♦	♦	♦	♦	♦	♦
Before proceeding, do homework: literature, suppliers, experts	♦	♦	♦	♦	♦	♦	♦	♦	♦	♦	♦

Index

toxic accumulation and, 29
water quality degradation and, 14–20
Biofouling bacteria, 158, see also
    specific types
Biological monitoring, 81–83
Bladder pumps, 60
Blended Chemical Heat Treatment
    (BCHT), 107, 149, 157
Blended method treatments, 149–150
Borehole video cameras, 76, 78
Budget analysis, 75
Budgeting, 111, see also Costs

C

Calcium, 79
Calcium carbonate, 8
Calcium sulfate, 8, 160–161
Case studies, 153–162
    Belgrade, Serbia, 155
    biofouling, 154–155, 156
    cleaning system, 156–157
    contaminated groundwater, 161–162
    environmental well system, 157–162
    extraction well, 161–162
    geotechnical, 155–157
    iron biofouling, 154–155
    long-term dewatering, 155–156
    monitoring well, 159–160
    New South Wales, Australia, 156–157
    northwestern Ontario, 154–155
    plume-control well, 158
    pump-and-treat system, 157–158
    recovery well, 158
    United Kingdom, 156
Casing, 43–45
Cement grouting, 44
Centrifugals, 60
Chemical encrustation, 6, 8
Chlorination, 30, 53, 107, 146–147
Chlorine, 30, 146, 147
Chlorine-based disinfectants, 30
Citric acid, 145
Clay, 9
Clay infiltration, 6
Cleaning chemicals, 114–115, see also
    specific types
Cleanliness practice, 54
Clogging, 7, 12, 14–15, 30, 66, see also
    Biofouling; Encrustation
Clogging precipitates, 31–32

Clostridium spp., 29
Cold carbon dioxide fracking, 140
Communication, 163–164
Conductivity, 82
Construction of wells, see Well
    construction
Continuing education, 3
Contractors, 123–128, 132–133, 137, 153
Conventional redevelopment, 139
Corrosion, 6, 8, 18, 105
    control of, 106
    costs of, 30
    intergranular, 21
    metallic, 20–21
    prevention of, 42–43
    resistance to, 42, 44
Cost-benefit analysis, 65, 109–115
Cost-effectiveness of maintenance, 66,
    113–114
Costs
    of chemicals, 30, 32–33
    of clogging, 30
    of corrosion, 30
    of doing nothing, 130–131
    of encrustation, 30
    of maintenance monitoring, 111–112
    of materials, 32–33
    operating, 32
    overhead, 121
    personnel, 65
    of preventive maintenance, 65, 69
    of preventive treatments, 109–115
    projections of, 113
    of rehabilitation, 119–120, 128–133
    for rehabilitation contractors,
        132–133
    types of, 30–32
    of well deterioration, 30–33
    of well performance, 30–33
    of well rehabilitation, 65
Culturing methods, 84–87, see also
    BART methods
Cyclohexane, 44

D

DBPs, see Disinfection by-products
"Designing for failure," 46
Design of wells, see Well design
Development of wells, see Well
    development